Diagnosis and Fault-tolerant Control 1

SCIENCES

Systems and Industrial Engineering,
Field Director – Jean-Paul Bourrières

Reliability, Diagnosis, Safety and Maintenance of Systems,
Subject Head – Jean-Marie Flaus

Diagnosis and Fault-tolerant Control 1

Data-driven and Model-based Fault Diagnosis Techniques

Coordinated by
Vicenç Puig
Silvio Simani

iSTE

WILEY

First published 2021 in Great Britain and the United States by ISTE Ltd and John Wiley & Sons, Inc.

ISTE Ltd
27-37 St George's Road
London SW19 4EU
UK

www.iste.co.uk

John Wiley & Sons, Inc.
111 River Street
Hoboken, NJ 07030
USA

www.wiley.com

Library of Congress Control Number: 2021943898

British Library Cataloguing-in-Publication Data
A CIP record for this book is available from the British Library
ISBN 978-1-78945-058-3

ERC Code:
PE7 Systems and Communication Engineering
PE7_1 Control engineering

Contents

Introduction ix
Vicenç PUIG and Silvio SIMANI

Chapter 1. Mathematical Modeling and Fault Description 1
Silvio SIMANI

1.1. Introduction 1
1.2. Model-based FDI techniques 2
1.3. Modeling of faulty systems 3
1.3.1. Fault modeling and description 5
1.3.2. Mathematical description 6
1.4. Residual generation 11
1.5. Residual generation techniques 14
1.5.1. Residual generation via parameter estimation 15
1.5.2. Observer-based approaches 18
1.5.3. Fault detection via parity equations 24
1.6. Change detection and symptom evaluation 28
1.7. Residual generation robustness problem 30
1.7.1. FDI H_∞ approach 32
1.7.2. Active and passive disturbance decoupling 35
1.8. Fault diagnosis technique integration 36
1.8.1. Fuzzy logic for residual generation 37
1.8.2. Neural networks for fault diagnosis 38
1.8.3. Neuro-fuzzy approaches to FDI 40
1.8.4. Fault detectability and isolability 42

1.8.5. NF model structure identification 43
1.8.6. NF residual generation for FDI 44
1.9. Conclusion 46
1.10. References 47

Chapter 2. Structural Analysis 57
Mattias KRYSANDER and Erik FRISK

2.1. Introduction 57
2.2. Background 58
2.2.1. Structural models 58
2.2.2. Dulmage–Mendelsohn decomposition and matchings 60
2.2.3. Dulmage–Mendelsohn decomposition and simulation 63
2.3. Fault isolability analysis 64
2.3.1. Fault detectability analysis 64
2.3.2. Fault isolability analysis 65
2.3.3. Canonical isolability decomposition of the overdetermined part . 67
2.4. Testable submodels 69
2.4.1. Basic definitions 69
2.4.2. MSO algorithm 71
2.4.3. Residual generation based on matching 72
2.5. Sensor placement 74
2.5.1. The basic sensor placement problem 74
2.5.2. A structural approach 75
2.6. Summary and discussion 80
2.7. References 81

Chapter 3. Set-based Fault Detection and Isolation 83
Ye WANG and Vicenç PUIG

3.1. Introduction 83
3.2. Notations, definitions and properties 84
3.3. Problem statement 86
3.3.1. Uncertain discrete-time linear systems 86
3.3.2. Set-based methods 86
3.3.3. FDI problem statement 88
3.4. Proposed techniques 89
3.4.1. Set-membership approach 89
3.4.2. Zonotopic observer 90
3.4.3. Relationship between set-based methods 91
3.5. Design methods 92
3.5.1. Robustness conditions 92
3.5.2. Fault sensitivity condition 96

3.6. Fault detection and isolation procedures 99
3.6.1. Fault detection 99
3.6.2. Fault isolation 100
3.7. Application example: quadruple-tank system 101
3.7.1. Results with robustness condition 105
3.7.2. Results with robustness and fault sensitivity conditions 105
3.8. Conclusion 105
3.9. References 109

Chapter 4. Diagnosis of Stochastic Systems 111
Gregory PROVAN

4.1. Introduction 111
4.2. Stochastic diagnosis task 113
4.2.1. Notation 113
4.2.2. Problem formulation 113
4.2.3. Representing uncertainty 115
4.3. Inference methods for diagnosis task 116
4.3.1. Difference with other tasks 116
4.4. Model-based approach 117
4.4.1. Traditional FDD methods 117
4.4.2. Bayesian inversion/filtering 120
4.5. Data-driven approaches 122
4.5.1. ML methods 123
4.5.2. Statistical methods 124
4.6. Hybrid approaches: surrogate methods 125
4.6.1. Fitting surrogate models via sampling 125
4.7. Comparative analysis of approaches 126
4.8. Summary and conclusions 127
4.9. References 128

Chapter 5. Data-Driven Methods for Fault Diagnosis 131
Silvio SIMANI

5.1. Introduction 131
5.2. Models for linear system fault diagnosis 133
5.3. Parameter estimation methods for fault diagnosis 135
5.3.1. Data-driven method in ideal conditions 135
5.3.2. Data-driven methods in real scenarios 138
5.3.3. Algebraic Frisch scheme 139
5.3.4. Dynamic Frisch scheme 141
5.3.5. MIMO case Frisch scheme 145
5.4. Nonlinear dynamic system identification 146
5.4.1. Piecewise affine model 147
5.4.2. Hybrid model structure 148
5.4.3. Nonlinear system approximation 149

5.4.4. Model continuity and domain partitioning 151
5.4.5. Local affine model estimation 154
5.4.6. Multiple-model estimation 158
5.5. Fuzzy data-driven approach to fault diagnosis 164
5.5.1. Fuzzy model identification 165
5.5.2. Takagi–Sugeno prototypes 167
5.5.3. Data-driven fuzzy modeling 170
5.5.4. Clustering methods 170
5.5.5. Fuzzy c-means clustering algorithms 172
5.5.6. Gustafson–Kessel clustering algorithm 174
5.5.7. Optimal number of clusters 176
5.6. Fuzzy model identification 176
5.6.1. Nonlinear model identification 178
5.6.2. Product space clustering identification 181
5.6.3. Fuzzy clustering model identification 183
5.6.4. Antecedent membership function estimation 183
5.6.5. Estimating consequent parameters 185
5.7. Conclusion 189
5.8. References 189

Chapter 6. The Artificial Intelligence Approach to Model-based Diagnosis 197
Belarmino PULIDO, Carlos J. ALONSO-GONZÁLEZ and Anibal BREGON

6.1. Introduction 197
6.2. Case studies 199
6.3. Knowledge-based diagnosis systems 201
6.3.1. Diagnosis task and system model 203
6.3.2. Diagnosis of physical devices 206
6.3.3. Limits of KBS for diagnosis of physical devices 207
6.4. Model-based diagnosis 208
6.4.1. Formalization of consistency-based diagnosis and its first implementation, GDE 209
6.5. CBD for dynamic systems 217
6.5.1. Different approaches for CBD of dynamic systems 219
6.5.2. PCs for the three-tank system case study 222
6.6. Conclusion 224
6.7. References 226

List of Authors 231

Index 233

Summary of Volume 2 237

Introduction

Vicenç PUIG[1] and Silvio SIMANI[2]
[1] *Institute for Robotics and Industrial Informatics, Polytechnic University of Catalonia, Barcelona, Spain*
[2] *Department of Engineering, University of Ferrara, Emilia-Romagna, Italy*

I.1. Introduction

There is an increasing interest in the theory and applications of model-based fault detection and fault diagnosis methods, because of economical and safety-related matters. In particular, well-established theoretical developments can be seen in many contributions published in the IFAC (International Federation of Automatic Control) Congresses and IFAC Symposium SAFEPROCESS (Fault Detection, Supervision and Safety of Technical Processes) since the seminal editions (Isermann and Ballé 1997; Isermann 1997; Patton 1999; Frank *et al.* 2000).

The developments began at various places in the early 1970s. Beard (1971) and Jones (1973) reported, for example, the well-known "failure detection filter" approach for linear systems.

A summary of this early development is given by Willsky (1976). Then, Rault *et al.* (1971) considered the application of identification methods to the fault detection of jet engines. Correlation methods were applied to leak detection (Siebert and Isermann 1976).

The first book on model-based methods for fault detection and diagnosis with specific application to chemical processes was published by Himmelblau (1978).

Diagnosis and Fault-tolerant Control 1,
coordinated by Vicenç PUIG and Silvio SIMANI. © ISTE Ltd 2021.

Sensor failure detection based on the inherent analytical redundancy of multiple observers was shown by Clark (1978).

The use of parameter estimation techniques for fault detection of technical systems was demonstrated by Hohmann (1977), Bakiotis *et al.* (1979), Geiger (1982) and Filbert and Metzger (1982).

The development of process fault detection methods based on modeling, parameter and state estimation was then summarized by Isermann (1984, 1997).

Parity equation-based methods were developed initially by Chow and Willsky (1984) and then further developed by Patton and Chen (1994), Gertler (1991) and Höfling and Pfeufer (1994).

Frequency domain methods are typically applied when the effects of faults as well as disturbances have frequency characteristics that differ from each other and thus the frequency spectra serve as criterion to distinguish the faults (Massoumnia *et al.* 1989; Ding and Frank 1990; Ding *et al.* 2000; Frank *et al.* 2000).

The developments of fault detection and isolation (FDI) methods to the present time are summarized in books by Pau (1981), Patton *et al.* (2000), Basseville and Nikiforov (1993), Chen and Patton (1999), Gertler (1998) and Isermann (1994b) as well as in survey papers by Gertler (1988), Frank (1990) and Isermann (1994a).

Within IFAC, the increasing interest in this field was taken into account by creating first in 1991 a SAFEPROCESS (Fault Detection Supervision and Safety for Technical Processes) Steering Committee, which then became a Technical Committee in 1993. The first IFAC SAFEPROCESS Symposium was held in Baden–Baden (Germany) in 1991 (Isermann and Freyermuth 1992). The last edition was organized in 2018 in Warsaw (Poland). Another triennial series of IFAC workshops exist for "Fault detection and supervision in the chemical process industries". Many other thematic workshops have been organized between 1992 and 2020.

Most contributions in fault diagnosis rely on the analytical redundancy principle. The basic idea consists of using an accurate model of the system to mimic the real process behavior. If a fault occurs, the residual signal (i.e. the difference between the real system and model behavior) can be used to diagnose and isolate the malfunction.

Model-based method reliability, which also includes false alarm rejection, is strictly related to the "quality" of the model and measurements exploited for fault diagnosis, as model uncertainty and noisy data can prevent an effective application of analytical redundancy methods.

This is not a simple problem, because model-based fault diagnosis methods are designed to detect any discrepancy between the real system and model behaviors. It is assumed that this discrepancy signal is related to (has a response from) a fault. However, the same difference signal can respond to model mismatch or noise in real measurements, which are erroneously detected as a fault. These considerations have led to research in the field of "robust" methods, in which particular attention is paid to the discrimination between actual faults and errors due to model mismatch.

However, the availability of a "good" model of the monitored system can significantly improve the performance of diagnostic tools, minimizing the probability of false alarms.

This book explains what a good model is, one that is suitable for robust diagnosis of system performance and operation. The book also explains how robust models can be obtained from real data. A large amount of attention is paid to the "real system modeling problem", with reference to either linear or nonlinear model structures. Special treatment is given to the case in which noise affects the acquired data. The mathematical description of the monitored system is obtained by means of a system identification scheme based on equation error and errors-in-variables models. This is an identification approach that leads to a reliable model of the plant under investigation, as well as the estimation of the variances of the input–output noises affecting the data.

The purpose of this book is also to provide guidelines for the modeling and identification of real processes for fault diagnosis and fault-tolerant control (FTC). Hence, significant attention is paid to the practical application of the methods describing real system studies, as reported in the last chapters of Volume 2.

In particular, this introduction of the book outlines a new common terminology in the fault diagnosis framework and provides some discussion and a summary of developments in the field of fault detection and diagnosis as well as FTC based on papers selected during 1991–2020.

I.2. Nomenclature

By going through the literature, one immediately recognizes that the terminology in this field is not consistent. This makes it difficult to understand the goals of the contributions and to compare the different approaches.

The IFAC SAFEPROCESS Technical Committee therefore discussed this matter and tried to find commonly accepted definitions. Some basic definitions can be found, for example, in the RAM (Reliability, Availability and Maintainability) dictionary (Omdahl 1988) and in contributions to the IFIP (International Federation for Information Processing) (IFI 1983).

Some of the terminology used in this book is given below. These are based on information obtained from the IFAC SAFEPROCESS Technical Committee and are considered "on-going" in the sense that new definitions and updates are still being made.

1) ***States and signals***

- **Fault**: an unpermitted deviation of at least one characteristic property or parameter of the system from the acceptable, usual or standard condition.

- **Failure**: a permanent interruption of a system's ability to perform a required function under specified operating conditions.

- **Malfunction**: an intermittent irregularity in the fulfillment of a system's desired function.

- **Error**: a deviation between a measured or computed value of an output variable and its true or theoretically correct value.

- **Disturbance**: an unknown and uncontrolled input acting on a system.

- **Residual**: a fault indicator based on a deviation between measurements and model-equation-based computations.

- **Symptom**: a change of an observable quantity from normal behavior.

2) ***Functions***

- **Fault detection**: determination of faults present in a system and the time of detection.

- **Fault isolation**: determination of the kind, location and time of detection of a fault. It follows fault detection.

- **Fault identification**: determination of the size and time-variant behavior of a fault. It follows fault isolation.

- **Fault diagnosis**: determination of the kind, size, location and time of detection of a fault. It follows that fault diagnosis includes fault detection and identification.

- **Monitoring**: a continuous real-time task of determining the conditions of a physical system by recording information, recognizing and indicating anomalies in the behavior.

- **Supervision**: monitoring a physical system and taking appropriate actions to maintain the operation in the case of a fault.

3) ***Models***

- **Quantitative model**: use of static and dynamic relationships among system variables and parameters in order to describe a system's behavior in quantitative mathematical terms.

- **Qualitative model**: use of static and dynamic relationships among system variables in order to describe a system's behavior in qualitative terms such as causalities and IF–THEN rules.

- **Diagnostic model**: a set of static or dynamic relationships that link specific input variables, the *symptoms*, to specific output variables, the *faults*.

- **Analytical redundancy**: use of more (not necessarily identical) ways to determine a variable, where one way uses a mathematical process model in an analytical form.

4) ***System properties***

- **Reliability**: ability of a system to perform a required function under stated conditions, within a given scope, during a given period of time.

- **Safety**: ability of a system to operate without causing danger to persons, equipment or the environment.

- **Availability**: probability that a system or equipment will operate satisfactorily and effectively at any point of time.

5) ***Time dependency of faults***

- **Abrupt fault**: fault modeled as step-wise function. It represents bias in the monitored signal.

- **Incipient fault**: fault modeled by using ramp signals. It represents drift of the monitored signal.

- **Intermittent fault**: combination of impulses with different amplitudes.

6) ***Fault terminology***

- **Additive fault**: it influences a variable by an addition of the fault itself. It may represent, for example, offsets of sensors.

- **Multiplicative fault**: it is represented by the product of a variable with the fault itself. It can appear as parameter changes within a process.

I.3. Fault diagnosis methods based on analytical redundancy

A traditional approach to fault diagnosis in the wider application context is based on *hardware* or *physical redundancy* methods, which use multiple sensors, actuators and components to measure and control a particular variable. Typically, a voting technique is applied to the hardware redundant system to decide if a fault has occurred and its location among all the redundant system components. The major problems encountered with hardware redundancy are the extra equipment and maintenance costs, as well as the additional space required to accommodate the equipment (Isermann 1997; Isermann and Ballé 1997).

In view of the conflict between reliability and the cost of adding more hardware, it is possible to use the dissimilar measured values together to cross-compare with each other rather than replicating each hardware individually. This is the meaning of *analytical* or *functional redundancy*. It exploits redundant analytical relationships among various measured variables of the monitored process (Patton *et al.* 1989; Chen and Patton 1999). Figure I.1 illustrates the concepts of hardware and analytical redundancy.

In the analytical redundancy scheme, the resulting difference generated from the comparison of different variables is called a *residual* or *symptom signal*. The residual should be zero when the system is in normal operation and should be different from zero when a fault has occurred. This property of the residual is used to determine whether faults have occurred (Patton *et al.* 1989; Chen and Patton 1999).

Consistency checking in analytical redundancy is normally achieved through a comparison between a measured signal and estimated values. The estimation is generated by a mathematical model of the plant considered. The comparison is done using the residual quantities that are computed as differences between the measured signals and the corresponding signals generated by the mathematical model (Patton *et al.* 1989; Chen and Patton 1999).

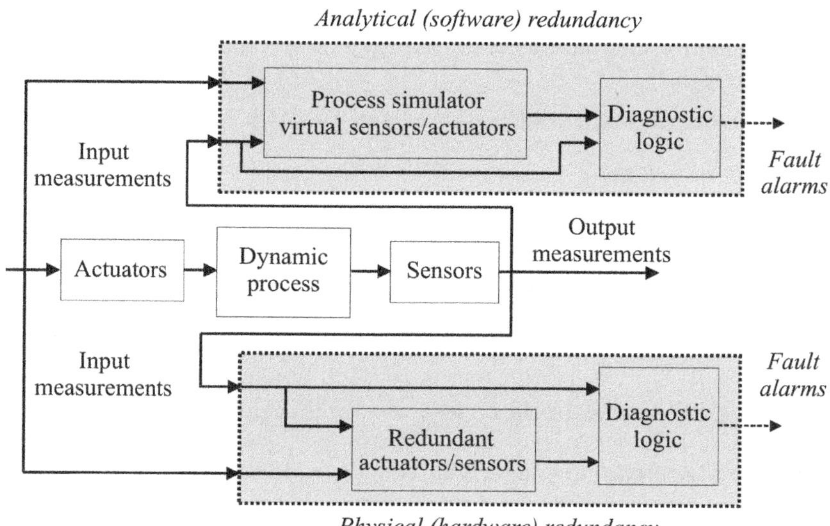

Figure I.1. *Comparison between hardware and analytical redundancy schemes*

In practice, the most frequently used diagnosis method is to monitor the level (or trend) of the residual and take action when the signal reaches a given threshold. This method of *geometrical analysis*, while simple to implement, has a few drawbacks. The most important is that, in the presence of noise, input variations and change of operating point of the monitored process, false alarms are possible.

The major advantage of the model-based approach is that no additional hardware components are required to implement an FDI algorithm as well as FTC. A model-based FDI algorithm can be implemented via software on a process control computer. In many cases, the measurements necessary to control the process are also sufficient for the FDI algorithm, so no additional sensors have to be installed (Patton *et al.* 1989; Basseville and Nikiforov 1993; Chen and Patton 1999).

Analytical redundancy uses a mathematical model of the system under investigation and therefore it is often referred to as the *model-based approach* to fault diagnosis.

I.4. Model-based fault diagnosis

This diagnosis task detects faults in the technical process, including actuators, components and sensors by measuring the available input and output variables $u(t)$ and $y(t)$. The principle of model-based fault diagnosis is depicted in Figure I.2.

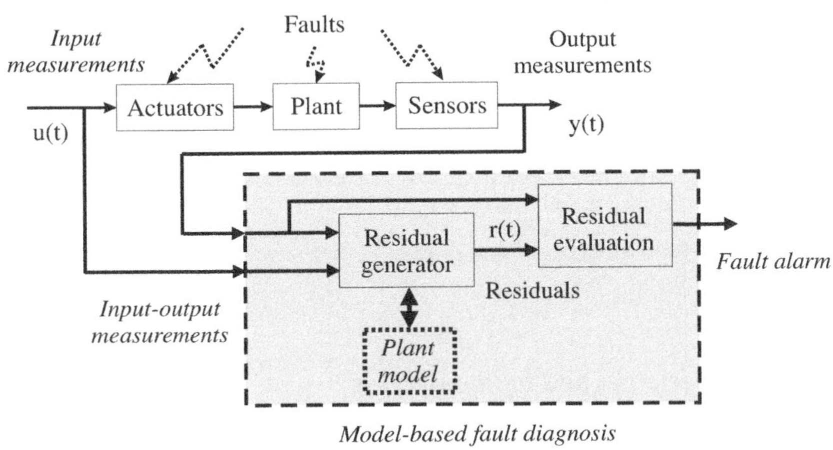

Figure I.2. *Scheme for the model-based fault diagnosis*

Basic process model-based FDI methods have been described by Patton *et al.* (1989, 2000), Basseville and Nikiforov (1993), Gertler (1998) and Chen and Patton (1999), which include the following steps:

1) output observers (OO, estimators, filters);

2) parity equations;

3) identification and parameter estimation.

These methods generate residuals for output variables with fixed parametric models using step 1, fixed parametric or non-parametric models using step 2 and adaptive non-parametric or parametric models using step 3.

An important aspect of these methods is the kind of fault to be detected. As noted above, one can distinguish between *additive faults*, which influence the variables of the process by summation, and *multiplicative faults*, which are products of the process variables. The basic methods show different results, depending on the type of fault.

If only output signals $y(t)$ can be measured, *signal model-based methods* can be applied, for example, vibrations can be detected, which are related to rotating machinery or electrical circuits. Typical signal model-based methods of fault detection are as follows:

1) bandpass filters;

2) spectral analysis (FFT);

3) maximum-entropy estimation.

The characteristic quantities or features from fault detection methods show stochastic behavior with mean values and variances. Deviations from the normal behavior must then be detected by methods of *change detection* (residual analysis, Figure I.2), such as:

1) mean and variance estimation;

2) likelihood-ratio test, Bayes decision;

3) run-sum test.

I.5. Model uncertainty and fault detection

Model-based FDI makes use of mathematical models of the system. However, a perfectly accurate mathematical model of a physical system is not possible. Usually, the parameters of the system may vary with time, and the characteristics of the disturbances and noises are unknown, so they cannot be modeled accurately. Hence, there is always a mismatch between the actual process and its mathematical model, even under no fault conditions. Such discrepancies cause difficulties in FDI applications, in particular, since they act as sources of false alarms and missed alarms. Therefore, the effect of modeling uncertainties, disturbances and noise is the most crucial point in the model-based FDI concept, and the solution to this problem is the key for its practical applicability (Chen and Patton 1999).

To overcome these problems, a model-based FDI scheme has to be insensitive to modeling uncertainty. Sometimes, a reduction of the sensitivity to modeling uncertainty does not solve the problem, because the sensitivity reduction may be associated with a reduction of the sensitivity to faults (Gertler 1998; Chen and Patton 1999). A more meaningful formulation of the FDI problem is to increase the insensitivity to modeling uncertainty in order to provide increasing fault sensitivity.

The difficulties introduced by model uncertainties, disturbances and noises in model-based FDI have been widely considered during the last 10 years by both academia and industry (Gertler 1998). A number of methods have been proposed to tackle this problem, for example, the Unknown Input Observer (UIO), eigenstructure assignment and parity relation methods.

An important task of the model-based FDI scheme is to be able to diagnose *incipient faults* in a system. With respect to *abrupt faults*, incipient faults may have a small effect on residuals and can be hidden by disturbances. On the other hand, hard faults can be detected more easily because their effects are usually larger than modeling uncertainties and a simple fixed threshold is usually enough to diagnose their occurrence by residual analysis.

The presence of incipient faults may not necessarily degrade the performance of the plant, however, they may indicate that the component should be replaced before the probability of more serious malfunctions increases. The successful detection and diagnosis of incipient faults can therefore be considered a challenge for the design and evaluation of FDI algorithms.

I.6. Robust fault diagnosis

In the context of automatic control, the term robustness is used to describe the insensitivity or invariance of the performance of control systems with respect to disturbances, model–plant mismatches or parameter variations. Fault diagnosis schemes, on the other hand, must, of course, also be robust to the mentioned disturbances, but, in contrast to automatic control systems, they must not be robust to actual faults. On the contrary, while generating robustness to disturbances, the designer must maintain or even enhance the sensitivity of fault diagnosis schemes to faults. Furthermore, the robustness, as well as the sensitivity properties, must be independent of the particular fault and disturbance mode. Generally, the problem of robust FDI can be divided into the tasks of *robust residual generation* followed by *robust residual evaluation*.

In many cases, the disturbances and model–plant mismatches to which robustness must be generated are due to the use of linear models for describing dynamic behavior of nonlinear processes. Modeling errors can be avoided from the very beginning by focusing on robust residual generation methods using linear and nonlinear process models. This, in turn, simplifies the problem of residual evaluation without reducing the sensitivity to actual faults.

Effective tools for robust residual generation and even complete decoupling from external disturbances and unknown system parameters can be provided, for example, by UIOs, which are introduced and applied to industrial processes. It is shown that the proposed solution to the disturbance decoupling problem also provides the solution to both the fault detection and fault isolation problems.

On the other hand, many dynamic processes can only be described effectively using nonlinear mathematical models. However, most of the existing observer-based FDI techniques are limited to the use of linear process models. The methods that can be found in the literature are based on the assumption that the system under supervision stays, during normal operation, in a neighborhood of a certain known operating point (Chen and Patton 1999; Patton *et al.* 2000)

It is clear that, as almost every process system is nonlinear, the modeling errors almost always reduce the accuracy of the linear model and therefore the performance of the FDI algorithm is compromised. Various methods for generating robustness to linearization have been proposed in the literature and the reader is referred to (Patton *et al.* 2000, Chapter 7) for a comprehensive treatment of this subject.

This book also surveys the state of the art of robustness methods and presents some important ideas concerning the development of the use of nonlinear models and predictors for FDI. For example, observer-based approaches to robust FDI and FTC for dynamic systems are considered in more detail. The available model-based approaches are generalized, and thus extended to a wider class of dynamic systems.

In order to accommodate the application of robust FDI concepts, disturbances and parameter uncertainties of the monitored plants, as well as faults, are modeled in the form of unknown input signals. It is shown that, provided certain conditions can be met, complete decoupling of the residual from disturbances as well as from the parameter uncertainties of the process model can be achieved, while the sensitivity of the residual to faults is maintained. As the faults are also modeled in the form of external signals, this method additionally provides tools for the purpose of fault isolation. Fault isolation requires the decoupling of the effects of different faults from the residual (Chen and Patton 1999) and this, in turn, allows for decisions on which fault or faults out of a given set of possible faults has actually occurred.

These residual properties must be completely independent of the magnitude or frequency of the unknown inputs and the faults. This is crucial in cases where no *a priori* knowledge about these properties is available. For systems where the complete decoupling of the remaining unknown inputs or faults from the residual proves impossible, a threshold selection method, employing functional analytic methods and appropriate vector and operator norms, can be exploited. This technique provides a tool for the robust evaluation of the residuals, which have been generated by UIOs. Using the same functional analysis methods as employed for threshold selection, a performance index can be defined that allows for performance evaluation and, to a certain degree, also allows for optimal residual generator design (Patton *et al.* 2000).

I.7. Data–driven approaches to robust FDI

In previous sections, we have seen that model-based FDI methods formally require a high accuracy mathematical model of the monitored system. The better the model is as a representation of the dynamic behavior of the system, the better the FDI performance will be. It is difficult to develop a highly accurate model of a complex system and hence the interesting question is: "what is a reasonable model to enable good performance in FDI to be guaranteed?".

It would be attractive to develop a robust FDI technique which is insensitive to modeling uncertainty, that is, so that a highly accurate mathematical model is no longer required. However, in order to design a robust FDI scheme, we should have a description (i.e. some information) about the uncertainty, for example, its *distribution matrix* and spectral bandwidth, and so on. Furthermore, this description should provide assistance for a robust FDI design, that is, it can be handled in a systematic manner. This book will show how a typical uncertainty description makes use of the concept of "unknown inputs" acting upon a nominal linear model of the system. These unknown disturbances describe the uncertainties acting upon the system but disturbance distribution matrices are assumed to be known since they can be estimated by identification schemes.

It is clear that disturbances and faults act on the system in the same way, and thus we cannot easily discriminate between these excitation signals unless we know the structure of the disturbance distribution matrix. Once the disturbance distribution matrix is known, we can generate the residual with the disturbance decoupling (robust) property, that is, the residual is decoupled from the disturbance (uncertainty). The robust residual can then be used to achieve reliable FDI and FTC.

The theories underlying robust FDI approaches have been very well developed, but for real applications the following problems remain unsolved:

– estimation of the reliable model for the monitored process;

– modeling accuracy of the real uncertainty by means of identified disturbance terms when no knowledge of the uncertainty is available;

– estimation of the disturbance terms and the structure of distribution matrices.

This book addresses these unsolved problems. Some simulation and real examples are given to test some of the theoretical results. These problems have to be addressed, otherwise the application domain of the disturbance decoupling approach for robust FDI is very limited. In fact, few researchers and contributions have presented the application results of robust fault diagnosis to real processes.

As mentioned above, a primary requirement for model-based and disturbance decoupling approaches to robust FDI is that both the system model and disturbance distribution matrices must be known. It is interesting that, within the framework of international research on this subject, there have been few attempts to address the problem by means of the *identification approach*. This lack of information has obstructed the application of robust FDI in real engineering systems. Therefore, we present the research developments surrounding the joint estimation of system and disturbance matrices in order to solve the robust fault diagnosis problem.

Concerning the data-driven schemes developed and exploited throughout the book, when all observed variables of a dynamic process are affected by uncertainties, the parameter estimation task can be performed by the so-called *errors-in-variables* methods. On the other hand, *equation error* methods can be developed in the case of exactly known plant variables (Simani *et al.* 2000). It is worthwhile noting that less attention has been paid to errors-in-variables schemes.

Under these considerations, this book presents the robust FDI results concerning the description of monitored plants by means of equation error and error-in-variables identified models in the presence of variable uncertainties. Moreover, for the examples presented, estimates obtained by the proposed data-driven approaches and parameter estimates will be computed and compared.

I.8. Data-driven methods for fault diagnosis

If several symptoms change differently for certain faults, an initial way of determining them is to use classification methods which indicate changes of symptom vectors.

Some classification methods are as follows (Patton *et al.* 1989; Basseville and Nikiforov 1993; Babuška 1998; Gertler 1998; Chen and Patton 1999):

1) geometrical distance and probabilistic methods;

2) artificial neural networks;

3) fuzzy clustering.

When more information about the relations between symptoms and faults is available in the form of diagnostic models, methods of reasoning can be applied. Diagnostic models then exist in the form of symptom–fault causalities, for example, in the form of symptom–fault trees. The causalities can be expressed as IF–THEN rules.

Then analytical as well as heuristic symptoms (from operators) can be processed. By considering these symptoms as vague facts, probabilistic or fuzzy set descriptions lead to a unified symptom representation. By using forward and backward reasoning, probabilities or possibilities of faults are obtained as a result of diagnosis. Typical approximate reasoning methods are as follows (Basseville and Nikiforov 1993; Chen and Patton 1999):

1) probabilistic reasoning;

2) possibilistic reasoning with fuzzy logic;

3) reasoning with artificial neural networks.

This very short consideration shows that many different methods have been developed over the last 30 years. It is also clear that many combinations of them are possible.

On the basis of different contributions during the last 30 years, it can be stated that parameter estimation and observer-based methods are the most frequently applied techniques for fault detection, especially for the detection of sensor and process faults. Nevertheless, the importance of neural network-based and combined methods for fault detection is steadily growing. In most applications, fault detection is supported by simple threshold logic or hypothesis testing. Fault isolation is often carried out using classification methods. For this task, neural networks are being more and more widely used.

The number of applications using nonlinear models is growing, while the trend of using linearized models is diminishing. It seems that analytical redundancy-based methods have their best application areas in mechanical systems where the models of the processes are relatively precise. Most nonlinear processes under investigation belong to the group of thermal and fluid dynamic processes. The field of applications to chemical processes has few developments, but the number of applications is growing. The favorite linear process under investigation is the DC motor. In general, the trend is changing from applications to safety-related processes with many measurements, as in nuclear reactors or aerospace systems, to applications in common technical processes with only a few sensors. For diagnosis, classification and rule-based reasoning methods are the most important, and the use of neural network classification as well as fuzzy logic-based reasoning is growing.

I.9. FDI application report

Because of the many publications and increasing number of applications (IFAC Congress and IFAC Symposia SAFEPROCESS) between 1991 and 2018, it is of interest to show some trends (Patton *et al.* 1989; Basseville and Nikiforov 1993; Gertler 1998; Chen and Patton 1999; Frank *et al.* 2000). Therefore, a literature study is briefly presented as follows. Contributions taking into account the applications reported in Table I.1 were considered. The type of faults considered is distinguished according to Table I.2. Among all contributions, the fault detection methods were classified as in Table I.3. The change detection and fault classification methods are indicated in Table I.4. The reasoning strategies for fault diagnosis are reported in Table I.5. The contributions considered are summarized in Table I.6. The evaluation has been limited to the fault detection and diagnosis (FDD) of laboratory, pilot and industrial processes.

Application	Number of contributions
Simulation of real processes	105
Large-scale pilot processes	94
Small-scale laboratory processes	68
Full-scale industrial processes	98

Table I.1. *FDI applications and number of contributions*

Fault type	Number of contributions
Sensor faults	129
Actuator faults	111
Process faults	123
Control loop or controller faults	48

Table I.2. *Fault type and number of contributions*

Method type	Number of contributions
Observer	123
Parity space	74
Parameter estimation	101
Frequency spectral analysis	57
Neural networks	79

Table I.3. *FDI methods and number of contributions*

Evaluation method	Number of contributions
Neural networks	89
Fuzzy logic	65
Bayes classification	54
Hypothesis testing	48

Table I.4. *Residual evaluation methods and number of contributions*

Reasoning strategy	Number of contributions
Rule based	40
Sign directed graph	33
Fault symptom tree	32
Fuzzy logic	66

Table I.5. *Reasoning strategies and number of contributions*

FDD	Number of contributions
Milling and grinding processes	91
Power plants and thermal processes	106
Fluid dynamic processes	67
Combustion engine and turbines	96
Automotive	68
Inverted pendulum	63
Miscellaneous	102
DC motors	121
Stirred tank reactor	77
Navigation system	75
Nuclear process	50

Table I.6. *Applications of model-based fault detection*

Table I.6 shows that among mechanical and electrical processes, DC motor applications are mostly investigated. Parameter estimation and observer-based methods are used in the majority of applications in these kind of processes, followed by parity space and combined methods. Thermal and chemical processes are investigated less frequently.

Table I.3 shows that parameter estimation and observer-based methods are used in nearly 70% of all applications considered. Neural networks, parity space and combined methods are applied notably less often.

More than 50% of sensor faults are detected using observer-based methods, while parameter estimation, parity space and combined methods play a less important role. For the detection of actuator faults, observer-based methods are mostly used, followed by parameter estimation and neural network methods.

Parity space and combined methods are rarely applied. In general, there are fewer applications for actuator faults than for sensor or process faults. The detection of process faults is mostly carried out with parameter estimation methods. Nearly 50% of all the applications considered use parameter estimation-based methods for detection of process faults. Observer-based, parity space and neural network-based methods are used less often for this class of faults.

Among all the described processes, linear models have been used much more than nonlinear models. In processes with nonlinear models, observer-based methods are mostly applied, but parity equations and neural networks also play an important role. In processes with linear or linearized models, parameter estimation and observer-based methods are mostly used. Parity space and combined methods are also used in several applications but not to the same extent as observer-based and parameter estimation methods.

Taking into account the system considered, the number of nonlinear process applications using nonlinear models is decreasing. For linear processes, no significant change can be stated. The applications of fault-detection methods for nonlinear processes used mostly observer-based and parameter estimation, more than parity space methods. Also, the use of neural networks and combinations are important.

Concerning the fault diagnosis methods, in recent years, the field of classification approaches, especially with neural networks and fuzzy logic, has steadily been growing. Also, rule-based reasoning methods are increasingly being based on fault diagnosis. A growing application of fuzzy rule-based reasoning can be stated. Applications using neural networks for classification are increasing and the trends are analogous to the increasing number of nonlinear process investigations. Nevertheless, the classification of generated residuals seems to remain the most important application area for neural networks.

I.10. From FDI to FTC

A conventional feedback control design for complex systems may result in unsatisfactory performance in the event of malfunction in input–output sensors, actuators and system components. A fault-tolerant closed-loop control system is very attractive because it can tolerate faults while also maintaining desirable performance.

The conventional approach to the design of an FTC includes different steps and separate modules: modeling or identification of the controlled system, design of the controller, FDI scheme and a method for re-configuring the control system. Identification and design of the controller can be performed separately or using combined methods. Hence, the FDI and controller are linked through the reconfiguration module. The fundamental problem with such a system lies in the identification stage in the independent design of the control and FDI modules. Significant interactions occurring among these modules can be neglected. There is therefore a need for a research study into the interactions between system identification, control design, the FDI stage and the FTC design strategy.

Fault identification is the most important of all the fault diagnosis tasks. When a fault is estimated, detection and isolation can be easily achieved since the fault nature can improve the diagnosis process. However, the fault identification problem itself has not gained enough research attention.

Most fault diagnosis techniques, such as parameter identification, parity space and observer-based methods, cannot be directly used to identify faults in sensors and actuators. Very little research has been done to overcome the fault identification problem. The Kalman filter for statistical testing and fault identification was proposed in Chen and Patton (1999). However, the statistical testing methods can impose a high computational demand. A fault identification scheme solving a system inversion problem was proposed in Chen and Patton (1999); Simani *et al.* (2003) and Simani and Farsoni (2018).

In the scheme, depicted in Figure I.3, fault identification is performed by estimating the nonlinear relationship between residuals and fault magnitudes. This is possible because robust residuals should only contain fault information.

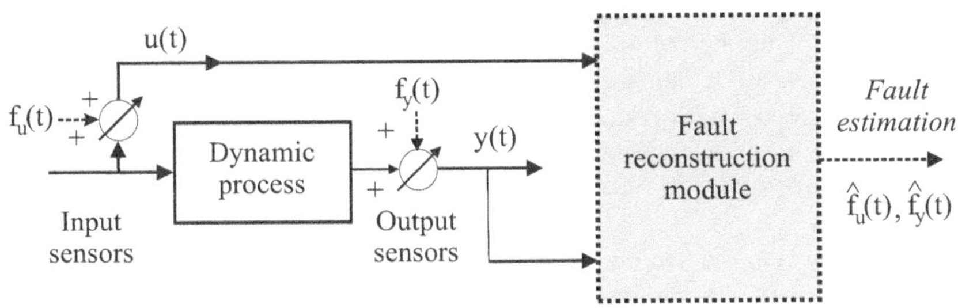

Figure I.3. *Fault estimation scheme FTC*

Such a nonlinear function approximation and estimation can be performed by using fuzzy systems, neural networks or an inversion of the transfer matrix between residuals and faults (Simani *et al.* 2003; Simani and Farsoni 2018). The central task in model-based fault detection is the residual generation. Most residual generation techniques are based on linear system models. For nonlinear systems, the traditional approach is to linearize the model around the system operating point. However, for systems with high nonlinearity and a wide dynamic operating range, the linearized approach fails to give satisfactory results.

One solution is to use a large number of linearized models corresponding to a range of operating points. This means that a large number of FDI schemes corresponding to each of the operating points is needed. Hence, it is important to study residual generation techniques that tackle nonlinear dynamic systems directly. There are some research studies on the residual generation of nonlinear dynamic systems, for example using nonlinear observers (Chen and Patton 1999; De Persis and Isidori 2001). There have been some attempts to use nonlinear observers to solve nonlinear system FDI problems, (Chen and Patton 1999; De Persis and Isidori 2001), for example, nonlinear UIOs, including adaptive observers and sliding mode observers. If the class of nonlinearities can be restricted, observers for bilinear systems were also proposed (Chen and Patton 1999).

On the other hand, the analytical models that the nonlinear observer approaches are based on are not easy to obtain in practice. Sometimes, it is impossible to model the system using an explicit mathematical model. To overcome this problem, it is desirable to find a universal approximate model that can be used to represent the real system with an arbitrary degree of accuracy. Different approaches were proposed and they are currently under investigation: neural networks, fuzzy models and hybrid models.

As shown in Simani *et al.* (2003) and Simani and Farsoni (2018), fuzzy systems and neural networks are a powerful tool for handling nonlinear problems. One of the most important advantages of neural networks is their ability to implement nonlinear transformations for functional approximation problems. Therefore, neural networks can be used in a number of ways to tackle fault diagnosis problems for nonlinear dynamic systems. In existing publications, they were mainly exploited as fault classifier with steady-state processes, whereas neural networks have been used as residual generators and for modeling nonlinear dynamic systems for FDI purposes (Chen and Patton 1999).

Fuzzy models can be used both as a residual classifier and as a nonlinear system parametric model (Chen and Patton 1999). In the second case, the main idea is to build an FDI scheme based on fuzzy observers. Estimated outputs and residuals are computed as fuzzy fusion of local observer output and residuals. The main problem of this approach concerns the stability of the global observer. A linear matrix inequality method was proposed in Chen and Patton (1999) using Lyapunov theorem, but this solution can be quite conservative.

Hybrid models can describe the behavior of any nonlinear dynamic process if they are described as a composition of several local affine models, selected according to the process operating conditions (Chen and Patton 1999; Simani *et al.* 2003). Instead of exploiting complicated nonlinear models obtained by modeling techniques, it is possible to describe the plant by a collection of affine models. Such a compound system requires the identification of the local models from data. Several works (Chen and Patton 1999; Simani *et al.* 2003) addressed a method for the identification and the optimal selection of the local affine models from a sequence of noisy measurements acquired from the process. Application of these results to model-based fault diagnosis for safety critical systems is another research area worthy of mention.

I.11. Structure of the book

To detect and isolate faults in a dynamic system, based on the use of an analytical model, a residual signal has to be used. It is derived from a comparison between real measurements and the relative estimates (generated by the model). The modeling uncertainty problem can be tackled by designing FDI and FTC schemes, whose residuals are insensitive to uncertainties while sensitive to faults. On the other hand, a model with satisfactory accuracy can be estimated using identification procedures (Norton 1986; Söderström and Stoica 1987; Ljung 1999).

The aim of the design of FDI and FTC schemes is to reduce the effects of uncertainties on the residuals and to enhance the effects of faults acting on the residuals. The *main aim of this book* is to develop a residual generator for model-based fault diagnosis and to design an effective FTC strategy for a dynamic process by means of input and output signals. An accurate model of the process under investigation will be estimated using identification procedures from data affected by noises and acquired from simulated and/or actual plants. The book consists of an Introduction and six chapters in Volume 1 and eight chapters in Volume 2 and the main contributions are summarized in the following.

The Introduction, provides a brief overview and critical discussion of the state of the art of the most recent literature from 2015 to 2020, thereby introducing the field of fault detection, fault diagnosis and fault-tolerant systems with methods, which have proven their significance in practical applications.

Chapter 1 addresses the mathematical modeling and description of the faults most commonly exploited for providing a proper description of the process under diagnosis, in connection with the strategy proposed for the diagnosis and FTC designs.

By taking into account these aspects, Chapter 2 is focused on structural analysis issues. In particular, this chapter addresses the standard tool used to identify submodels that can be used to design model-based and data-driven diagnostic modules. Structural approaches typically operate on models described by a set of equations, which can also be obtained from model-free approaches.

With reference to FDI, Chapter 3 considers set-based methods. The set-membership and interval observer approaches are introduced to deal with the robustness problem in fault detection. The design conditions to guarantee robustness, and at the same time fault sensitivity, are presented. Next, the extension to fault isolation using unknown-input observer schemes is described.

Chapter 4 describes stochastic methods for FDI. In particular, the chapter revises the existing methods for FDI using stochastic modeling of uncertainty, using both models and data.

As an alternative to analytical approaches, Chapter 5 proposes data-driven schemes. It is therefore devoted to the problem of fault detection in technical systems described by nonlinear dynamical models containing non-smooth nonlinearities. The so-called "model-free" or "data-driven" solutions can be exploited to solve the considered FTD and FTC problems. The feature of this method is that parameters of the system under consideration may be unknown.

Among data-driven solutions, Chapter 6 considers the artificial intelligence (AI) approach to fault diagnosis. After revising the evolution of fault diagnosis methods in the AI domain, the chapter focuses on the model-based approach rooted in the logic theory of diagnosis.

When considering analytical approaches, Chapter 1 of Volume 2 proposes the development of nonlinear methods. This chapter gives a review of the principal model-based fault diagnosis and fault-tolerant approaches for nonlinear systems. Some schemes extending the well-known diagnosis methods for linear systems to the nonlinear case are considered. The robustness of these schemes in the presence of uncertainty is discussed. Similarities between the approaches considered are also pointed out.

With reference to Volume 2, the problem of the FTC is addressed. In particular, Chapter 2 of Volume 2 considers the use of linear parameter varying (LPV) methods. In particular, this chapter considers FDI and FTC for descriptor LPV systems with unmeasurable decision variables under actuator faults and perturbations.

A different approach is considered in Chapter 3 of Volume 2, where fuzzy Takagi–Sugeno and neural network methods for FDI and FTC are revised. After introducing the different types of models, their application to fault diagnosis and estimation is presented. The extension to FTC is then described.

Chapter 4 of Volume 2 presents the model predictive control (MPC) techniques to deal with robustness and nonlinearity. To this aim, the use of neural networks is considered.

Chapter 5 of Volume 2 considers nonlinear methods for FTC. This chapter presents a methodology for detecting, isolating and accommodating faults in a class of nonlinear dynamic systems. On the basis of the fault information obtained by the fault-diagnosis procedure, an FTC component is designed to compensate for the effects of faults.

Chapter 6 of Volume 2 proposes virtual sensor and actuator development. The problem of FTC for dynamic processes is considered by using virtual sensor/actuator approaches to deal with sensor and actuator faults. This chapter also presents the extension to LPV systems using the Linear Matrix Inequality (LMI) approach.

Finally, Chapters 7 and 8 of Volume 2 complete the book by providing some concluding remarks and open research directions.

In particular, Chapter 7 of Volume 2 summarizes the main achievements of the book by highlighting the key features of the proposed diagnosis and fault-tolerant solutions when applied to safety critical systems.

Finally, Chapter 8 of Volume 2 analyzes some perspectives in the field of diagnosis and FTC by exploring open problems and future issues that could require further investigation. Future possible research directions are also outlined.

Therefore, the book reviews the state of the art of the data-driven and model-based FDI and FTC. The FDI and FTC problems are formalized in an uniform framework by presenting the mathematical description and definitions. The fundamental issue of model-based methods is the generation of residuals using the mathematical model of the monitored system. By analyzing residuals, fault diagnosis and FTC can be performed. Some structures of the residual generator are recalled to give ideas as to how to implement the residual generation. A residual generator can be designed for achieving the required diagnosis performances, for example, fault isolation and disturbance decoupling.

In order to design the residual generator, some assumptions about the modeling uncertainties need to be made. The most frequently used hypothesis is that the modeling uncertainty is expressed as a disturbance term in the system dynamic equation. The disturbance vector is unknown, while its distribution matrix can be estimated by using identification procedures. On the basis of this assumption, the disturbance decoupling residual generator can be designed by using UIO methods Chen and Patton (1999); Liu and Patton (1998).

The book also demonstrates how to apply dynamic system identification methods and more general data-driven approaches in order to estimate an accurate model of the monitored system.

The FDI and FTC methods presented can, in fact, require a mathematical model of the process under investigation, either in a state–space or an input–output form.

In particular, since state–space descriptions provide general and mathematically rigorous tools for system modeling, they may be used in the residual generator design, both for the deterministic case (generalized observers, such as UIOs and output dynamic observers) (Chen and Patton (1999); Frank (1990); Luenberger (1979); Watanabe and Himmelblau (1982)) and the stochastic case (such as Kalman filters and unknown input Kalman filters) (Jazwinski (1970); Xie *et al.* (1994); Xie and Soh (1994)).

In such a manner, the suggested FDI and FTC tools may not require any physical knowledge of the process under observation since the linear models are obtained by means of an identification scheme, which can, for example, exploit equation error (EE) and errors-in-variables (EIV) models. In this situation, the identification technique is based on the rules of the Frisch scheme (Frisch 1934), traditionally exploited to analyze economic systems. This approach, modified to be applied to dynamic system identification (Kalman 1982b, 1990; Beghelli *et al.* 1990), gives a reliable model of the plant under investigation, as well as the variances of the input–output noises affecting the data.

For the nonlinear case, fuzzy models and neural networks can be used as prototypes for the identification. In particular, the multiple-model approach, using several local affine submodels each describing a different operating condition of the process, is exploited.

Under these considerations, this book aims to define a comprehensive methodology for actuator, process component and sensor fault detection. It is based on an output estimation approach, in conjunction with residual processing schemes, which include simple threshold detection, in a deterministic case, as well as statistical analysis when data are affected by noise. The final result consists of a strategy based on fault diagnosis methods well known in the literature for generating redundant residuals.

In particular, this work studies different approaches to residual generation and fault compensation with the aid of several methodologies. In general, the residual is defined as the *output estimation error*, obtained by the difference between the measurement of one output and the relative estimate. This work also presents the design of such estimators both in the deterministic and stochastic environment.

The diagnosis procedure may be further specialized for actuators, input or output sensors and process components. In fact, the fault diagnosis of input sensors and actuators uses banks of estimators in high signal-to-noise ratio conditions, or filters, otherwise. The general principle designs the ith reconstructor to be insensitive to the ith signal of the system. On the other hand, output sensor and process component faults affecting a single residual can be detected by means of output observer or filters, driven by a single output and all the inputs of the system.

The book shows how the proposed algorithms can be applied to the FDI and FTC of actuators, process components and input–output sensors of industrial plants.

In particular, the different techniques presented in this book have been tested on time series of data acquired from different simulated and realistic industrial processes, energy conversion systems, power plants, and more general safety critical systems, whose linear mathematical description is obtained by using data-driven and model-based procedures.

Results from simulation show that minimum detectable faults are perfectly compatible with the industrial target of this application.

Finally, the book concludes the proposed research and application topics by summarizing its contributions and achievements, providing some suggestions for possible further research topics as an extension of this work.

I.12. Summary

This Introduction has provided a common terminology in the fault diagnosis framework in order to comment on some developments in the field of fault detection and diagnosis based on papers selected from the last 30 years. The structure of the 14 chapters and their main contributions have also been outlined briefly.

I.13. References

Babuška, R. (1998). *Fuzzy Modeling for Control*. Kluwer Academic Publishers. Boston, MA, USA.

Bakiotis, C., Raymond, J., Rault, A. (1979). Parameter identification and discriminant analysis for jet engine mechanical state diagnosis. *IEEE Conference on Decision and Control, Fort Lauderdale*, Fort Lauderdale, FL, USA.

Basseville, M. and Nikiforov, I.V. (1993). *Detection of Abrupt Changes: Theory and Application*. Prentice-Hall, Inc., Upper Saddle River, NJ, USA.

Beard, R.V. (1971). Failure accomodation in linear systems through self-reorganisation. Technical Report MVT-71-1, Man Vehicle Lab, Cambridge, MA, USA.

Beghelli, S., Guidorzi, R.P., Soverini, U. (1990). The Frisch scheme in dynamic system identification. *Automatica*, 26(1), 171–176.

Chen, J. and Patton, R.J. (1999). *Robust Model–Based Fault Diagnosis for Dynamic Systems*. Kluwer Academic Publishers, Boston, MA, USA.

Chow, E.Y. and Willsky, A.S. (1984). Analytical redundancy and the design of robust detection systems. *IEEE Transactions Automatic Control*, 29(7), 603–614.

Clark, R.N. (1978). Instrument fault detection. *IEEE Transactions on Aerospace & Electronic Systems*, 14(3), 456–465.

De Persis, C. and Isidori, A. (2001). A geometric approach to non-linear fault detection and isolation. *IEEE Transactions on Automatic Control*, 45(6), 853–865.

Ding, X. and Frank, P.M. (1990). Fault detection via factorization approach. *Syst. Contr. Lett.*, 14(5), 431–436.

Ding, S.X., Jeinsch, T., Frank, P.M., Dind, E.L. (2000). A unified approach to the optimisation of fault detection systems. *International Journal of Adaptive Control and Signal Processing*, 14(7), 725–745.

Filbert, D. and Metzger, K. (1982). Quality test of systems by parameter estimation. *9th IMEKO Congress*, Berlin, Germany.

Frank, P.M. (1990). Fault diagnosis in dynamic systems using analytical and knowledge based redundancy: A survey of some new results. *Automatica*, 26(3), 459–474.

Frank, P.M., Ding, S.X., Köpper-Seliger, B. (2000). Current developments in the theory of FDI. *SAFEPROCESS'00: Preprints of the IFAC Symposium on Fault Detection, Supervision and Safety for Technical Processes*, vol. 1, Budapest, Hungary, 16–27.

Geiger, G. (1982). Monitoring of an elecrical driven pump using continuous-time parameter estimation models. *6th IFAC Symposium on Identification and Parameter Estimation*, Washington, Pergamon Press, Oxford, UK.

Gertler, J. (1988). Survey of model-based failure detection and isolation in complex plants. *IEEE Control System Magazine*, 8(6), 3–11.

Gertler, J. (1991). Generating directional residuals with dynamic parity equations. *IFAC/IMACS Symposium Proceedings: SAFEPROCESS'91*, Baden-Baden, Germany.

Gertler, J. (1998). *Fault Detection and Diagnosis in Engineering Systems*. Marcel Dekker, New York, USA.

Himmelblau, D.M. (1978). *Fault Diagnosis in Chemical and Petrochemical Processes. Elsevier*, Amsterdam, The Netherlands.

Höfling, T. and Pfeufer, T. (1994). Detection of additive and multiplicative faults – Parity space vs. parameter estimation. *IFAC Symposium Proceedings: SAFEPROCESS'94*, Espoo, Finland.

Hohmann, H. (1977). Automatic monitoring and failure diagnosis for machine tools. Dissertation, T.H. Darmstadt, Germany [in German].

IFI (1983). Reliable computing and fault tolerance. Meeting, IFI, Como, Italy.

Isermann, R. (1984). Process fault detection based on modeling and estimation methods: A survey. *Automatica*, 20(4), 387–404.

Isermann, R. (1994a). Integration of fault detection and diagnosis methods. *IFAC Symposium Proceedings SAFEPROCESS'94*, Espoo, Finland.

Isermann, R. (1994b). Supervision and fault diagnosis. Paper, VDI-Verlag, Düsseldorf [in German].

Isermann, R. (1997). Supervision, fault detection and fault diagnosis methods: An introduction. *Control Engineering Practice*, 5(5), 639–652.

Isermann, R. and Ballé, P. (1997). Trends in the application of model-based fault detection and diagnosis of technical processes. *Control Engineering Practice*, 5(5), 709–719.

Isermann, R. and Freyermuth, B. (eds) (1992). *Fault Detection, Supervision and Safety for Technical Processes*. Pergamon Press, Oxford, UK.

Jazwinski, A.H. (1970). *Stochastic Processes and Filtering Theory*. Academic Press, New York, USA.

Jones, H.L. (1973). Failure detection in linear systems. PhD Thesis, Department of Aeronautics, M.I.T., Cambridge, MA, USA.

Kalman, R.E. (1982). System identification from noisy data. In *Dynamic Systems II, a University of Florida International Symposium*, Bednarek, A.R., Cesari, L. (eds). Academic Press, New York, USA.

Kalman, R.E. (1990). Nine lectures on identification. In *Lecture Notes on Economics and Mathematical System*, Fandel, G. and Trockel, W. (eds). Springer, Berlin, Germany.

Liu, G.P. and Patton, R.J. (1998). *Eigenstructure Assignment for Control System Design*. John Wiley & Sons, Chichester, UK.

Ljung, L. (1999). *System Identification: Theory for the User*, 2nd edition. Prentice Hall, Englewood Cliffs, N.J, USA.

Luenberger, D.G. (1979). *Introduction to Dynamic Systems: Theory, Models and Application*, John Wiley & Sons, New York, USA.

Massoumnia, M., Verghese, G.C., Willsky, A.S. (1989). Failure detection and identification. *IEEE Transactions on Automatic Control*, 34, 316–321.

Norton, J. (1986). *An Introduction to Identification*. Academic Press, London, UK.

Omdahl, T.P. (1988). *Reliability, Availability and Maintainability Dictionary*. ASQC Quality Press, Milwaukee, WI, USA.

Patton, R.J. (1999). Preface to the papers from the 3rd IFAC Symposium SAFEPROCESS'97. *Control Engineering Practice*, 7(1), 201–202.

Patton, R.J. and Chen, J. (1994). A review of parity space approaches to fault diagnosis for aerospace systems. *AIAA Journal of Guidance, Control & Dynamics*, 17(2), 278–285.

Patton, R.J., Frank, P.M., Clark, R.N. (eds) (1989). *Fault Diagnosis in Dynamic Systems, Theory and Application*, Prentice Hall, London, UK.

Patton, R.J., Frank, P.M., Clark, R.N. (eds) (2000). *Issues of Fault Diagnosis for Dynamic Systems*. Springer-Verlag, London, UK.

Pau, L.F. (1981). *Failure Diagnosis and Performance Control*. Marcel Dekker, New York, USA.

Rault, A., Richalet, A., Barbot, A., Sergenton, J.P. (1971). Identification and modelling of a jet engine. *IFAC Symposium od Digital Simulation of Continuous Processes*, Gejör.

Siebert, H. and Isermann, R. (1976). Fault diagnosis via online correlation analysis. Technical Report 25-3, VDI/VDE Darmstadt, Germany [In German].

Simani, S. and Farsoni, S. (2018). *Fault Diagnosis and Sustainable Control of Wind Turbines: Robust Data–driven and Model–based Strategies*, 1st edition. Butterworth–Heinemann, Elsevier, Oxford, UK.

Simani, S., Fantuzzi, C., Beghelli, S. (2000). Diagnosis techniques for sensor faults of industrial processes. *IEEE Transactions on Control Systems Technology*, 8(5), 848–855. DOI: 10.1109/87.865858.

Simani, S., Fantuzzi, C., Patton, R.J. (2003). *Fault Diagnosis in Dynamic Systems Using Identification Techniques*, 1st edition. Springer–Verlag, London, UK.

Watanabe, K. and Himmelblau, D.M. (1982). Instrument fault detection in systems with uncertainties. *International Journal of System Science*, 13(2), 137–158.

Willsky, A.S. (1976). A survey of design methods for failure detection in dynamic systems. *Automatica*, 12(6), 601–611.

Xie, L. and Soh, Y.C. (1994). Robust Kalman filtering for uncertain systems. *Systems and Control Letters*, 22, 123–129.

Xie, L., Soh, Y.C., de Souza, C.E. (1994). Robust Kalman filtering for uncertain discrete-time systems. *IEEE Transaction on Automatic Control*, 39, 1310–1314.

1

Mathematical Modeling and Fault Description

Silvio SIMANI
Department of Engineering, University of Ferrara, Emilia-Romagna, Italy

1.1. Introduction

The model-based approach to fault detection in dynamic systems has been receiving more and more attention over the last two decades, in the contexts of both research and real plant application.

Stemming from this activity, a large number of methods can be found in current literature based on the use of the mathematical models of the process under investigation and on exploiting modern control theory.

Model-based fault detection methods use residuals that indicate changes between the process and the model. One general assumption is that the residuals are changed significantly so that detection is possible. This means that the residual size after the appearance of a fault is large and long enough to be detectable.

This chapter provides an overview on the various fault detection methods, with particular attention to the fault detection and isolation (FDI) techniques related to the applications described in this book.

For all of the methods considered, it is essential that the process can be described by a mathematical model. As there is almost never an exact agreement between the

Diagnosis and Fault-tolerant Control 1,
coordinated by Vicenç PUIG and Silvio SIMANI.

model used to represent the process and the process itself, the model–reality discrepancy is of primary interest.

Hence, the most important issue in model-based fault detection concerns the accuracy of the model describing the behavior of the monitored system. This issue has become a central research theme over recent years, as modeling uncertainty has risen from the impossibility of obtaining complete knowledge and understanding of the monitored process.

The main focus of this chapter is the modeling aspects of the process whose faults are to be detected and isolated. The chapter also studies the general structure of a controlled system, its possible fault locations and modes. Residual generation is then identified as an essential problem in model-based FDI, because, if it is not performed correctly, some fault information could be lost. A general framework for the residual generation is also recalled.

Residual generators based on different methods, such as state and output observers, parity relations and parameter estimations, are just special cases in this general framework. In the following, some commonly used residual generation and evaluation methods are discussed and their mathematical formulation is presented.

Finally, the chapter presents and summarizes special features and problems regarding the different methods.

1.2. Model-based FDI Techniques

According to the definitions given in the literature, model-based FDI can be defined as the *detection*, *isolation* and *identification* of faults on a system by means of methods, which extract features from measured signals and use *a priori* information on the process available in terms of mathematical models.

Faults are, thus, detected by setting fixed or variable thresholds on residual signals generated from the difference between actual measurements and their estimates obtained by using the process model.

A number of residuals can be designed, with each having sensitivity to individual faults occurring in different locations of the system. The analysis of each residual, once the threshold is exceeded, then leads to fault isolation.

Figure 1.1 shows the general and logic block diagram of the model-based FDI system.

It comprises two main stages of residual generation and residual evaluation. This structure was first suggested by Chow and Willsky (1980) and now is widely accepted by the fault diagnosis community.

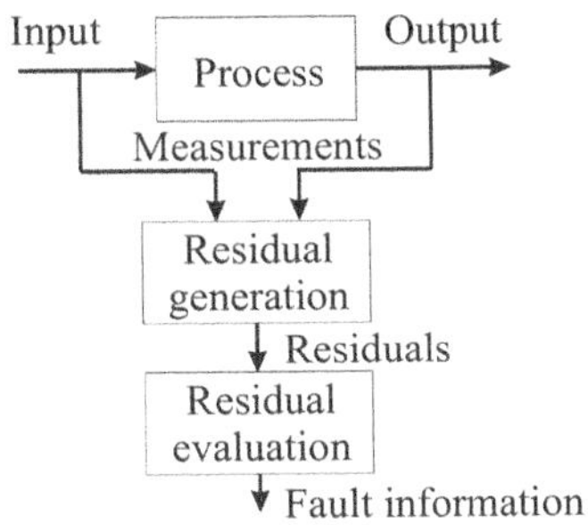

Figure 1.1. *Structure of the model-based FDI system*

The two main blocks are described as follows:

1) **Residual generation**: this block generates residual signals using available inputs and outputs from the monitored system. This residual (or fault symptom) should indicate that a fault has occurred. It should normally be zero or close to zero under no fault condition, while indistinguishably different from zero when a fault occurs. This means that the residual is characteristically independent of process inputs and outputs, in ideal conditions. Referring to Figure 1.1, this block is called *residual generation*.

2) **Residual evaluation**: this block examines residuals for the likelihood of faults and a decision rule is then applied to determine if any faults have occurred. The *residual evaluation* block, shown in Figure 1.1, may perform a simple threshold test (geometrical methods) on the instantaneous values or moving averages of the residuals. On the other hand, it may consist of statistical methods, for example, generalized likelihood ratio testing or sequential probability ratio testing (Isermann 1997; Willsky 1976; Basseville 1988; Patton *et al.* 2000).

Most contributions in the field of quantitative model-based FDI focus on the residual generation problem, since the decision-making problem can be considered relatively straightforward if residuals are well designed.

Section 1.3 presents a number of different strategies for solving the quantitative residual generation problem.

1.3. Modeling of faulty systems

This chapter is mainly concerned with multi-input single-output (MISO) and multi-input multi-output (MIMO) dynamic systems.

The first step in the FDI model-based approach consists of providing a mathematical description of the system under investigation, which also shows all of the possible fault cases.

The detailed scheme for the FDI techniques presented here is depicted in Figure 1.2.

The main components are the *Plant* under investigation, the *Actuators* and *Sensors*, which can be further sub-divided as *input* and *output* sensors, and finally, the *Controller*.

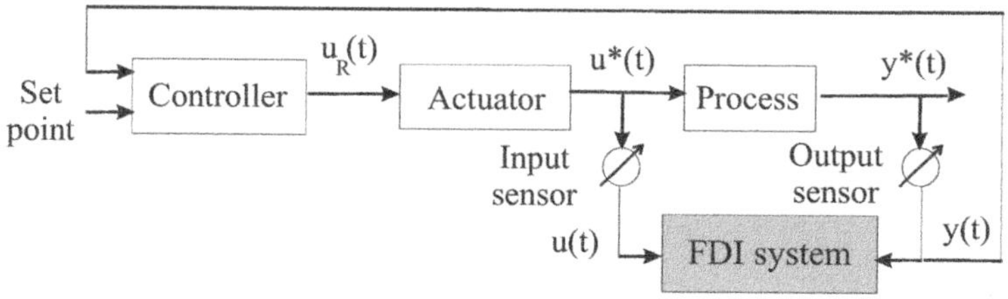

Figure 1.2. *Closed-loop FDI system*

In the following, the system working conditions will be monitored by means of its input $u(t)$ and output $y(t)$ measurements and signals from the controller $u_R(t)$, which are supposedly completely available for FDI purposes. Also, as shown in Figure 1.3, the behavior of any controller that drives the system is inherently taken into consideration.

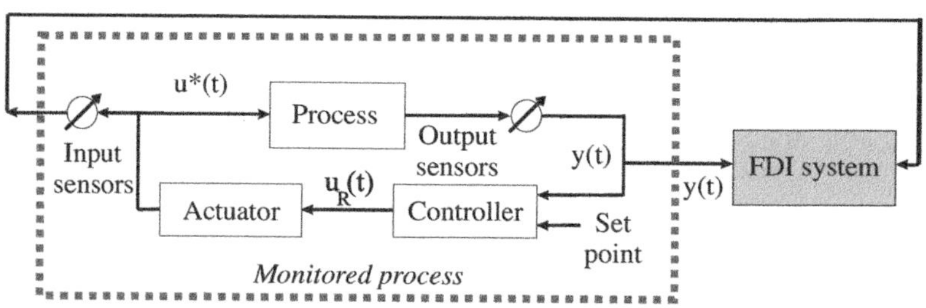

Figure 1.3. *The rearranged FDI scheme*

It is worth noting that, when the signals $u_R(t)$ from the controller or measurements of plant inputs $u(t)$ are not available, the controller plays an important role in the design of the FDI scheme, as a robust controller may desensitize faults effects and make diagnosis difficult.

Once the actual process inputs and outputs $u^*(t)$ and $y^*(t)$ (usually unavailable) are measured by the input and output sensors, FDI theory can be treated as an observation problem of $u(t)$ and $y(t)$. The monitored system considered for FDI purpose can be therefore rearranged, as illustrated in Figure 1.3.

1.3.1. *Fault modeling and description*

Concerning the occurrence of malfunctions, the *location of faults* and their modeling, the system under diagnosis can be separated into the following different parts, which can be affected by faults:

- actuators;
- process or system components;
- input sensors;
- output sensors;
- controllers.

With respect to previous work (see, e.g. Patton *et al.* 1989, 2000; Gertler 1998), it is necessary to distinguish between input and output sensors.

Figure 1.3 shows that the input and output signals $u^*(t)$ and $y^*(t)$ are acquired in order to obtain the measurements $u(t)$ and $y(t)$ from the sensors. This fault scenario can be summarized by the diagram shown in Figure 1.4.

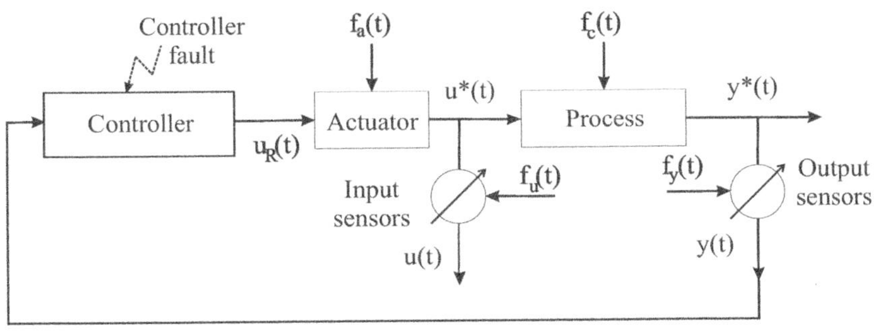

Figure 1.4. *Monitored system and fault topology*

Figure 1.4 also shows the situation where the controller can be affected by faults, since the monitored process consists of a closed-loop system. However, because of technological reasons (e.g. the control action is performed by a digital computer), when the actuator is considered as a part or a component of the whole controller device, the former can be treated as a subsystem where faults are likelier to occur, while the latter remains free from faults.

Under these assumptions, as depicted in Figure 1.5, when the system is considered in view of fault location, since input and output measurements are supposedly completely available for FDI purposes, the controller behavior in the design of a fault diagnosis scheme can be neglected, as well as the interconnection between the control system and the process.

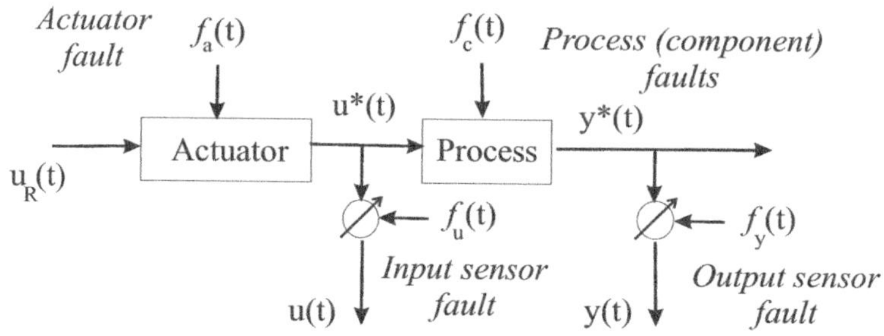

Figure 1.5. *The monitored system and fault typology*

1.3.2. *Mathematical description*

Under the hypothesis of linearity, the process dynamics considered in this chapter are described by the following discrete-time, time-invariant, linear dynamic system in the state–space form

$$\begin{cases} x(t+1) = Ax(t) + Bu^*(t) \\ y^*(t) \quad = Cx(t) \end{cases} \qquad [1.1]$$

where $x(t) \in \Re^n$ is the system state vector, $u^*(t) \in \Re^r$ is the input signal vector driven by actuators and $y^*(t) \in \Re^m$ is the real system output vector, which is not directly available. A, B and C are system matrices with appropriate dimensions obtained by a modeling or identification procedure.

With reference to Figure 1.5, a component fault vector $f_c(t)$ affects process dynamics as follows:

$$x(t+1) = Ax(t) + Bu^*(t) + f_c(t) \qquad [1.2]$$

In some cases, component faults come from a change in the system parameters, for example, a change in the entries of the A matrix. For example, a change in the ith row and the jth column of the A matrix leads to a fault vector $f_c(t)$ described as

$$f_c(t) = I_i \Delta a_{ij} x_j(t) \qquad [1.3]$$

where $x_j(t)$ is the jth element of the vector $x(t)$ and I_i is an n-dimensional vector with all zero, except a "1" in the ith element.

As stated previously, as the actual process output $y^*(t)$ is not directly available, a sensor is used to acquire a measure of the system outputs. Moreover, generally speaking, a sensor can also be used to measure the system inputs $u^*(t)$ (e.g. for an uncontrolled system).

By neglecting sensor dynamics, faults on input and output sensors are modeled with additive signals, respectively, as:

$$\begin{cases} u(t) = u^*(t) + f_u(t) \\ y(t) = y^*(t) + f_y(t) \end{cases} \qquad [1.4]$$

where the vectors $f_u(t) = [f_{u_1}(t) \dots f_{u_r}(t)]^T$ and $f_y(t) = [f_{y_1}(t) \dots f_{y_m}(t)]^T$ are chosen to describe a fault situation.

For example, if the sensor outputs are stuck at a fixed value $\bar{u}$ because of a malfunction, the measurement vector is $u(t) = \bar{u}$ and the fault can be written as $f_u(t) = -u^*(t) + \bar{u}$.

On the other hand, when the sensors are affected by a multiplicative fault δ, the measurements become $u(t) = (1 + \delta)u^*(t)$, and the fault vector can be written as $f_u(t) = \delta u^*(t)$.

Usually, as shown in the following, fault modes can be described by step and ramp signals in order to model abrupt and incipient (hard to detect) faults, representing bias and drift, respectively.

Moreover, for technical reasons, sensor output signals are generally affected by measurement noise. Fault-free sensor signals $u(t)$ and $y(t)$ with additive noise can be modeled as:

$$\begin{cases} u(t) = u^*(t) + \tilde{u}(t) \\ y(t) = y^*(t) + \tilde{y}(t) \end{cases} \qquad [1.5]$$

in which the sequences $\tilde{u}(t)$ and $\tilde{y}(t)$ are usually described as white, zero-mean, uncorrelated Gaussian processes.

In this case, taking into account the effects of faults and noise, [1.4] has to be replaced by:

$$\begin{cases} u(t) = u^*(t) + \tilde{u}(t) + f_u(t) \\ y(t) = y^*(t) + \tilde{y}(t) + f_y(t) \end{cases} \qquad [1.6]$$

By neglecting the actuator block, Figure 1.6 shows the structure of the measurement process.

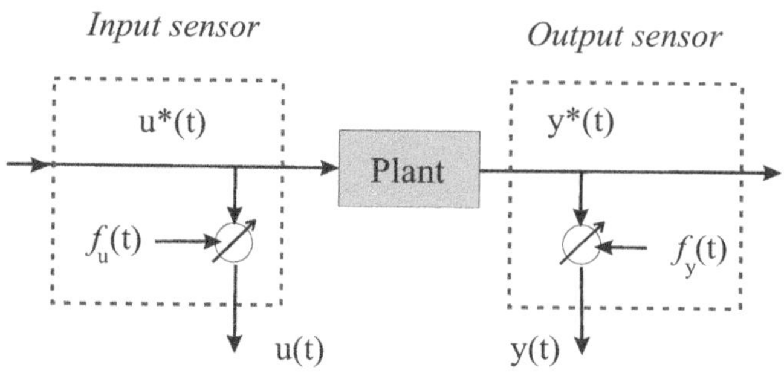

Figure 1.6. *The structure of the plant sensors*

Model descriptions of the types of equations [1.1] and [1.5] are also known as Errors-In-Variables models (Kalman 1982b, 1990). They will be briefly recalled later in this chapter.

With reference to a controlled system, according to Figure 1.5, signals $u^*(t)$ are the actuator response to the command signals $u_R(t)$.

A purely algebraic actuator (i.e. with gain equal to 1) can be described by:

$$u^*(t) = u_R(t) + f_a(t) \qquad [1.7]$$

where, similarly to the input–output sensor fault situation, $f_a(t) \in \Re^r$ is the actuator fault vector.

In general, as shown in Figure 1.5, if the actuation signals $u^*(t)$ are assumed to be measurable, by neglecting input and output sensor noises, the process model with fault can be described by the following system equation:

$$\begin{cases} x(t+1) = Ax(t) + f_c(t) + Bu^*(t) \\ y(t) \quad\;\; = Cx(t) + f_y(t) \\ u(t) \quad\;\; = u^*(t) + f_u(t) \end{cases} \qquad [1.8]$$

On the other hand, Figure 1.7 represents the case where the u_R signals are measured only by the input sensors.

Such a configuration represents a critical situation with respect to the input sensor connection depicted in Figure 1.5.

In this situation, actuator faults cannot be directly related to the input measurements $u(t)$, but their effects can only be detected by means of output signals $y(t)$.

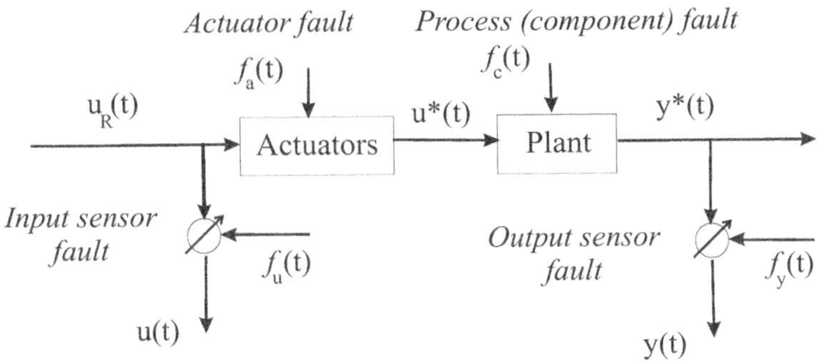

Figure 1.7. *Fault topology with actuator input signal measurement*

By also taking into account actuator faults $f_a(t)$, the below description is obtained:

$$\begin{cases} x(t+1) = Ax(t) + f_c(t) + Bf_a(t) + Bu^*(t) \\ y(t) \quad\; = Cx(t) + f_y(t) \\ u(t) \quad\; = u^*(t) + f_u(t) \end{cases} \quad [1.9]$$

Moreover, considering the general case, a system affected by all possible faults can be described by the following state–space model:

$$\begin{cases} x(t+1) = Ax(t) + Bu^*(t) + L_1 f(t) \\ y(t) \quad\; = Cx(t) + L_2 f(t) \\ u(t) \quad\; = u^*(t) + L_3 f(t) \end{cases} \quad [1.10]$$

where the entries of the vector $f(t) = [f_a^T, f_u^T, f_c^T, f_y^T]^T \in \Re^k$ correspond to specific faults.

In practice, it is reasonable to assume that the fault signals are described by *unknown* time functions. The matrices L_1, L_2, L_3 are known as faulty entry matrices, which describe how the faults enter the system.

The vectors $u(t)$ and $y(t)$ are the available and measurable inputs and outputs, respectively. Both vectors are supposedly known for FDI purposes.

The distribution of the fault in the system, depicted in Figure 1.5, can be described as an input–output transfer matrix representation in the following form:

$$y(z) = G_{yu^*}(z)u^*(z) + G_{yf}(z)f(z) \quad [1.11]$$

with z being the unitary advance operator while the transfer matrices $G_{yu^*}(z)$ and $G_{yf}(z)$ are defined as:

$$\begin{cases} G_{yu^*}(z) = C(zI - A)^{-1}B \\ G_{yf}(z) \ = C(zI - A)^{-1}L_1 + L_2 \end{cases} \qquad [1.12]$$

Both of the general models for FDI described by equations [1.10] and [1.11] in the time and frequency domain, respectively, have been widely accepted in the fault diagnosis literature (Patton *et al.* 1989, 2000; Chen and Patton 1999; Gertler 1998).

Under these assumptions, the general model-based FDI problem treated here can only be performed on the basis of the knowledge of the measured sequences $u(t)$ and $y(t)$.

Frequency domain descriptions are typically applied when the effects of faults, as well as the disturbances, have frequency characteristics which differ from each other, and thus, information in the frequency spectra serve as criteria to distinguish the faults (Ding and Frank 1990; Massoumnia *et al.* 1989).

On the other hand, since state–space descriptions provide general and mathematically rigorous tools for system modeling and robust residual generation, both for the deterministic case (noise free measurements) and the stochastic case (measurements affected by noises), the system matrices A, B and C, [1.10], in canonical forms can be obtained by multivariable identification procedures (Guidorzi 1975; Norton 1986; Söderström and Stoica 1987; Ljung 1999).

Moreover, in the case of a MIMO system, the choice of state–space representations in canonical form (Guidorzi 1975) instead of parity space methods (Gertler 1995) may avoid unexpected false alarm problems (Delmaire *et al.* 1999).

As known from the system identification framework, the FDI methods proposed here do not require any physical knowledge of the processes under observation, since the mathematical description of the monitored system is obtained by means of a system identification scheme based on Equation Error (EE) and EIV models.

It is worthy to note how this approach represents a novel point of view of the model-based fault diagnosis. The new aspect consists of exploiting linear system identification procedures, in connection with the model-based residual generation problem.

Although most systems to be monitored are actually nonlinear, linear system modeling and identification methods are described here to avoid the complexities that would otherwise be inevitable when nonlinear models are used.

There is certainly an increasing interest in the use of nonlinear methods (nonlinear observers, extended Kalman filters, fuzzy-logic methods, etc.). However, as the feature of system supervision is to monitor the operation and performance of the system with respect to an expected point of operation, linear system methods are still very valid. Deviations from expected behavior can be used to monitor system performance changes, as well as component malfunctions.

1.4. Residual generation

In this section, a review is given on fault detection methods based on *process models* and *signal models*. The basic methods are described briefly while their presentation and application are shown in this book.

The most frequently used FDI methods exploit the *a priori* knowledge of the characteristics of certain signals. As an example, the spectrum, the dynamic range of the signal and its variations may be checked. However, the necessity of *a priori* information concerning the monitored signals and the dependence of the signal characteristics on the unknown working conditions of the system under diagnosis are the main drawbacks of such a class of methods.

The most significant contribution in modern model-based approaches is the introduction of the *symptom or residual signals*, which depend on faults and are independent of system operating states.

They represent the inconsistency between the actual system measurements and the corresponding signals of the mathematical model.

The residual generator block introduced in Figure 1.1 can be interpreted as illustrated in Figure 1.8 (Basseville 1988).

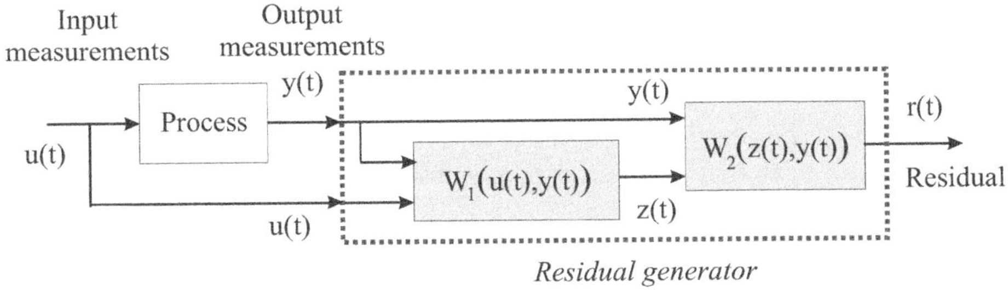

Figure 1.8. *Residual generator general structure*

In the above structure, the auxiliary redundant signal $z(t)$ is generated by the function $W_1\big(u(\cdot), y(\cdot)\big)$ and, together with the measurement $y(t)$, the symptom signal $r(t)$ is computed by means of $W_2\big(z(\cdot), y(\cdot)\big)$.

In the fault-free case, the following relations are satisfied:

$$\begin{cases} z(t) = W_1\big(u(\cdot), y(\cdot)\big) \\ r(t) = W_2\big(z(\cdot), y(\cdot)\big) = 0. \end{cases} \qquad [1.13]$$

When a fault occurs in the plant, the residual $r(t)$ will be different from zero.

The simplest residual generator is depicted in Figure 1.9 and it is obtained when the system W_1 is a plant identical model $z(t) = W_1\big(u(\cdot)\big)$, or it is an input–output description for the actual process obtained from the system identification procedure (e.g. an Auto-Regressive eXogenous (ARX) model).

In the former case, the measurement $y(t)$ is not required in W_1 because it is a *system simulator*. The signal $z(t)$ represents the simulated output and the residual is computed as $r(t) = z(t) - y(t)$. Since it is an open-loop system, the process simulation may become unstable.

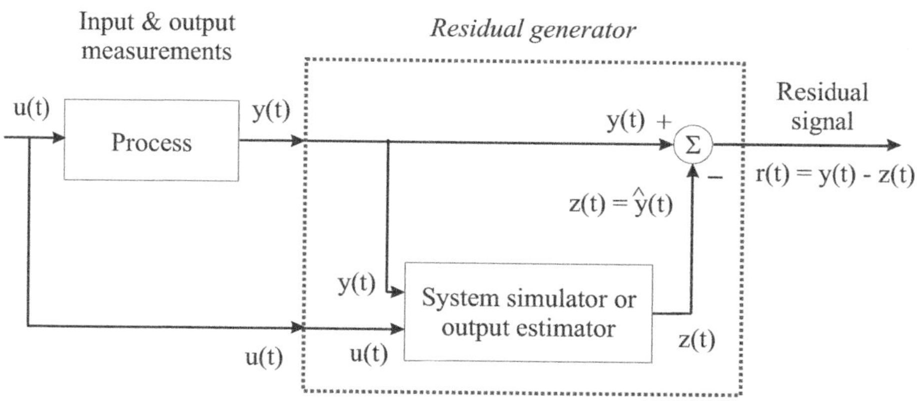

Figure 1.9. *System simulator or output estimator for residual generation*

An extension to the model-based residual generation is to replace $W_1\big(u(\cdot)\big)$ by $W_1\big(u(\cdot), y(\cdot)\big)$, that is, an *output estimator* fed by both system input and output.

In such a case, function W_1 generates an estimation of a linear function of the output $W_1\big(u(\cdot), y(\cdot)\big) = My(t)$, while function W_2 can be defined as $W_2\big(z(\cdot), y(\cdot)\big) = W\big(z(t) - My(t)\big)$, with W as a weighting matrix.

Therefore, no matter which type of method is used, the residual generation process is nothing but a liner mapping whose inputs consist of process inputs and outputs.

As an example, Figure 1.10 represents a general structure for all residual generators using the input–output transfer matrix description presented in Patton and Chen (1991a).

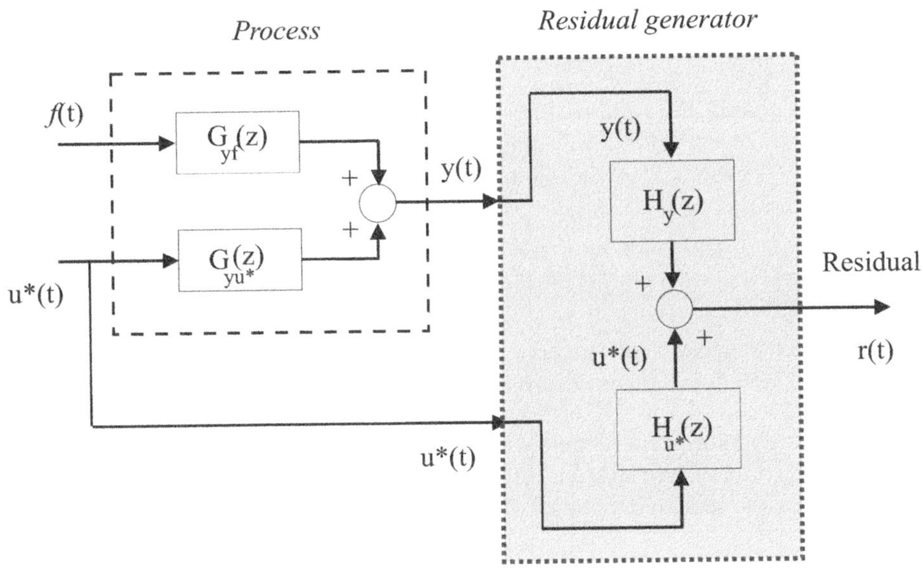

Figure 1.10. *Transfer function residual generator*

With reference to equations [1.11] and [1.12], the residual generator structure is expressed mathematically by the generalized representation:

$$r(z) = \begin{bmatrix} H_{u^*}(z) & H_y(z) \end{bmatrix} \begin{bmatrix} u^*(z) \\ y(z) \end{bmatrix} = H_{u^*}(z)u^*(z) + H_y(z)y(z) \qquad [1.14]$$

where $H_{u^*}(z)$ and $H_y(z)$ are discrete transfer matrices that can be designed using stable discrete-time linear systems. The functions $u^*(z)$, $y(z)$, $r(z)$ and $f(z)$ are the Z-transform of the corresponding discrete-time signals.

According to the definition, the residual $r(t)$ has to be designed to become zero for a fault-free case and different from zero in case of failures. This means that

$$r(t) = 0 \text{ if and only if } f(t) = 0 \qquad [1.15]$$

In order to satisfy equation [1.15], the design of the transfer matrices $H_{u^*}(z)$ and $H_y(z)$ must satisfy the constraint conditions

$$H_{u^*}(z) + H_y(z)G_{yu^*} = 0 \quad [1.16]$$

It is worth noting that different residual generators can be obtained by using different parameterizations of $H_{u^*}(z)$ and $H_y(z)$ (Patton and Chen 1991a; Chen and Patton 1999).

After generating the residual, the simplest and most widely used way to fault detection is achieved by directly comparing the residual signal $r(t)$ or the residual function $J(r(t))$ with a fixed threshold ϵ or a threshold function $\varepsilon(t)$ as follows:

$$\begin{cases} J(r(t)) \leq \varepsilon(t) \text{ for } f(t) = 0 \\ J(r(t)) > \varepsilon(t) \text{ for } f(t) \neq 0 \end{cases} \quad [1.17]$$

where $f(t)$ is the general fault vector defined in equation [1.10]. If the residual exceeds the threshold, a fault may have occurred.

This test works especially well with fixed thresholds ε if the process operates approximately in a steady state and it reacts after a relatively large feature, that is, after either a large sudden or a long-lasting gradually increasing fault.

On the other hand, adaptive thresholds $\varepsilon(t)$ that depend on plant operating conditions can be exploited, for example when $\varepsilon(t)$ is expressed as a function of plant inputs (Clark 1989; Chen and Patton 1999).

1.5. Residual generation techniques

The generation of symptoms is the main issue in model-based fault diagnosis.

A variety of methods are available in literature for residual generation and this section briefly presents some of the most common methods.

Most of the residual generation techniques are based on both continuous and discrete system models; however, in this book, the attention is focused on discrete-time dynamic linear models.

The following process model-based fault detection schemes will be considered and summarized (Isermann and Ballé 1997; Patton *et al.* 2000):

1) fault detection via parameter estimation (Isermann 1984, 1993; Isermann and Freyermuth 1992; Isermann and Ballé 1997; Patton *et al.* 2000);

2) observer-based approaches (Beard 1971; Frank 1993; Frank and Ding 1997; Patton and Chen 1997; Willsky 1976; Basseville 1988);

3) parity vector (relation) methods (Chow and Willsky 1984; Gertler and Singer 1990; Patton and Chen 1991a; Gertler and Monajemy 1993; Delmaire *et al.* 1999).

1.5.1. *Residual generation via parameter estimation*

In most practical cases, the process parameters are not known at all, or they are not known exactly enough. Then, they can be determined with parameter estimation methods, by measuring input and output signals $u(t)$ and $y(t)$, if the basic structure of the model is known (Isermann 1997; Patton *et al.* 2000).

This approach is based on the assumption that the faults are reflected in the physical system parameters and the basic idea is that the parameters of the actual process are estimated online using well-known parameter estimations methods.

The results are thus compared with the parameters of the reference model, obtained initially under fault-free assumptions. Any discrepancy can indicate that a fault may have occurred.

Now we compare two different approaches for modeling the input–output behavior of the monitored system.

1.5.1.1. *Equation error methods*

The SISO process discrete-time model of order n is written in the vector form:

$$y(t) = \Psi^T \Theta \quad [1.18]$$

where

$$\Theta^T = [a_1 \dots a_n,\ b_1 \dots b_n] \quad [1.19]$$

is the parameter vector, and

$$\Psi^T = [y(t-1) \dots y(t-n)\ \ u(t-1) \dots u(t-n)] \quad [1.20]$$

is the discrete-time data vector.

According to Figure 1.11, for parameter estimation, the EE $e(t)$ is introduced:

$$e(t) = y(t) - \Psi^T \Theta \quad [1.21]$$

or, if

$$\frac{y(t)}{u(t)} = \frac{B(z)}{A(z)} \qquad [1.22]$$

is the transfer function of the process, the EE via Z-transformation becomes

$$e(t) = \hat{B}(z)u(t) - \hat{A}(z)y(t) \qquad [1.23]$$

in which $\hat{A}(z)$ and $\hat{B}(z)$ correspond to the estimates of $A(z)$ and $B(z)$.

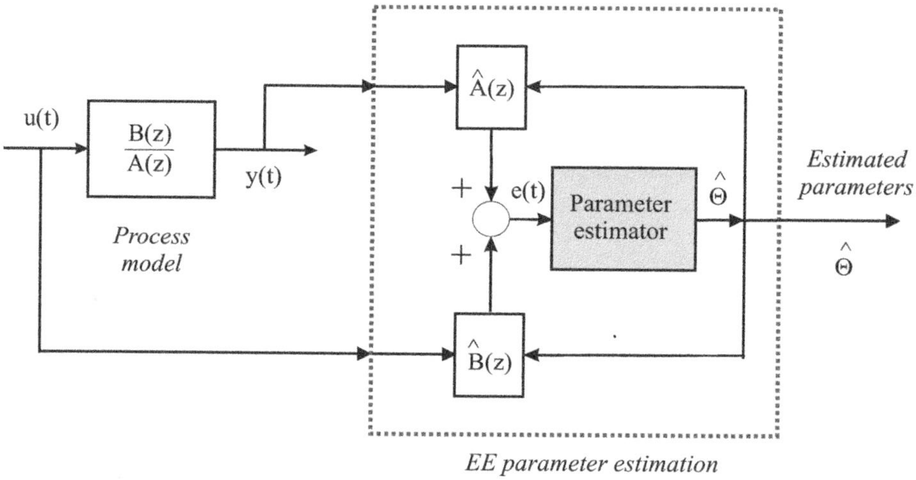

Figure 1.11. *Parameter estimation Equation Error (EE)*

The least-squares (LS) estimate

$$\hat{\Theta} = [\Psi^T \, \Psi]^{-1} \Psi^T y \qquad [1.24]$$

is obtained if the minimization of the sum of LS is computed

$$\begin{cases} J(\Theta) &= \sum_t e^2(t) = e^T e \\ \frac{d\, J(\Theta)}{d\, \Theta} &= \qquad 0. \end{cases} \qquad [1.25]$$

As described in Patton *et al.* (2000) and Isermann (1992), the LS estimate can also be computed via a recursive least square (RLS) algorithm in real-time form, with respect to the estimates at the instant t, with $t = 0,\ 1,\ 2, \cdots$:

$$\hat{\Theta}(t+1) = \hat{\Theta}(t) + \gamma(t)\left[y(t+1) - \Psi^T(t+1)\hat{\Theta}(t+1)\right] \quad [1.26]$$

where

$$\begin{cases} \gamma(t) & = \frac{1}{\Psi^T(t+1)P(t)\Psi(t+1)+1} P(t)\Psi(t+1) \\ P(t+1) = & \left[I - \gamma(t)\Psi^T(t+1)\right] P(t). \end{cases} \quad [1.27]$$

For improved estimates, filtering methods can be exploited. In particular, when measurements are affected by noise, a Kalman filter can be used for the parameter estimation (Jazwinski 1970).

1.5.1.2. *Output error methods*

Instead of the EE computed in equation [1.21], the Output Error (OE)

$$e(t) = y(t) - \hat{y}(\Theta, t) \quad [1.28]$$

where

$$\hat{y}(\Theta, z) = \frac{\hat{B}(z)}{\hat{A}(z)} u(z) \quad [1.29]$$

which is the model output, can also be used, as depicted in Figure 1.12.

Unfortunately, a direct calculation of the parameter estimate Θ is not possible, because $e(t)$ is nonlinear in the parameters.

Therefore, the loss function of equation [1.28] as equation [1.21] has to be minimized by numerical optimization methods. The computational effort is then much larger and online real-time application is generally impossible. However, relatively precise parameter estimates may be obtained.

If a fault within the process changes one or several parameters by $\Delta\Theta$, the output signal changes for small deviations according to

$$\Delta y(t) = \Psi^T(t)\Delta\Theta(t) + \Delta\Psi^T(t)\Theta(t) + \Delta\Psi^T(t)\Delta\Theta(t) \quad [1.30]$$

and the parameter estimator indicates a change $\Delta\Theta$.

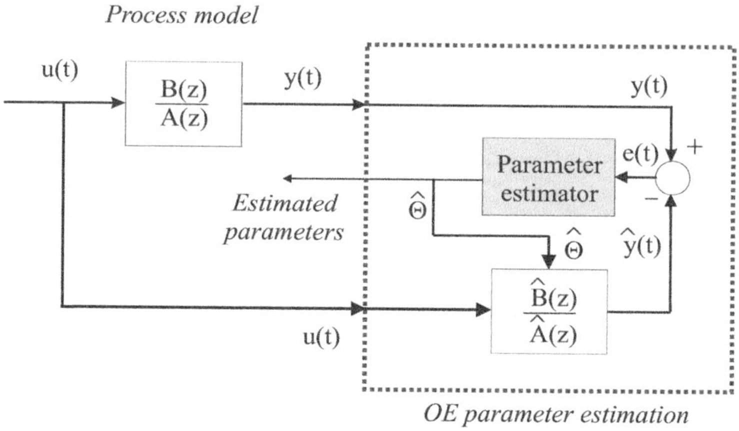

Figure 1.12. *Parameter estimation Output Error (OE)*

Generally, the process parameters Θ depend on physical process coefficients p (such as stiffness, damping factor and resistance):

$$\Theta = f(p) \quad [1.31]$$

via nonlinear algebraic equations. If the inversion of the relationship

$$p = f^{-1}(\Theta) \quad [1.32]$$

exists (Patton *et al.* 2000; Isermann 1992), changes Δp of the process coefficients can be calculated. These changes in the coefficients are in many cases, directly related to faults.

Thus, although the knowledge of Δp facilitates the fault diagnosis problem, it is not necessary for fault detection only. Parameter estimation can also be applied to nonlinear static process models (Isermann 1993).

1.5.2. *Observer-based approaches*

The basic idea behind the observer or filter-based techniques is to estimate the outputs of the system from the measurements by using either Luenberger observers in a deterministic setting, or Kalman filters in a noisy environment. The output estimation error (or its weighted value) is therefore used as residual.

It is worth noting that when an observer is exploited for FDI purpose, the estimation of the outputs is necessary, while the estimation of the state vector is usually not needed (Chen and Patton 1999). Moreover, the advantage of using the observer is the flexibility in the selection of its gains, which leads to a rich variety of FDI schemes (Frank 1994b; Frank and Ding 1997; Chen *et al.* 1996b; Liu and Patton 1998).

In order to obtain the structure of a (generalized) observer, the discrete-time, time-invariant, linear dynamic model for the process under consideration in a state–space form is considered:

$$\begin{cases} x(t+1) = Ax(t) + Bu(t) \\ y(t) \quad\;\; = Cx(t) \end{cases} \qquad [1.33]$$

being $u(t) \in \Re^r$, $x(t) \in \Re^n$ and $y(t) \in \Re^m$.

Assuming that all matrices A, B and C are perfectly known, an observer is used to reconstruct the system variables based on the measured inputs and outputs $u(t)$ and $y(t)$:

$$\begin{cases} \hat{x}(t+1) = A\hat{x}(t) + Bu(t) + He(t) \\ e(t) \quad\;\; = y(t) - C\hat{x}(t). \end{cases} \qquad [1.34]$$

The observer scheme described by equation [1.34] is depicted in Figure 1.13.

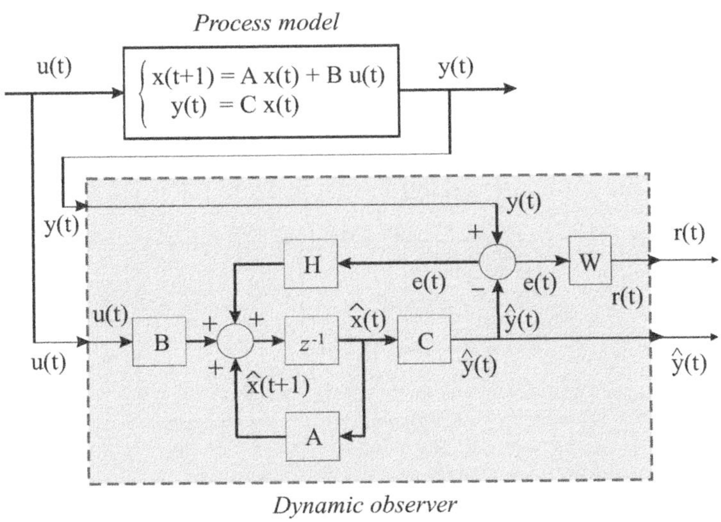

Figure 1.13. *State and output dynamic observer*

For the state estimation error $e_x(t)$, it follows from equations [1.34] that

$$\begin{cases} e_x(t) &= x(t) - \hat{x}(t) \\ e_x(t+1) &= (A - HC)e_x(t). \end{cases} \quad [1.35]$$

The state error $e_x(t)$ (and the error $e(t)$) vanishes asymptotically)

$$\lim_{t \to \infty} e_x(t) = 0 \quad [1.36]$$

if the observer is stable, which can be achieved by proper design of the observer feedback H.

If the process is influenced by disturbance and faults, by comparing Figure 1.14 and equations [1.10] it is described by the following system:

$$\begin{cases} x(t+1) &= Ax(t) + Bu(t) + Qv(t) + L_1 f(t) \\ y(t) &= Cx(t) + Rw(t) + L_2 f(t) \end{cases} \quad [1.37]$$

where $v(t)$ is the non-measurable disturbance vector at the input, $w(t)$ is the non-measurable disturbance vector at the output and$f(t)$ are the fault signals at the input and output, acting through L_1 and L_2, respectively.

They can represent actuator, process, input and output sensor additive faults.

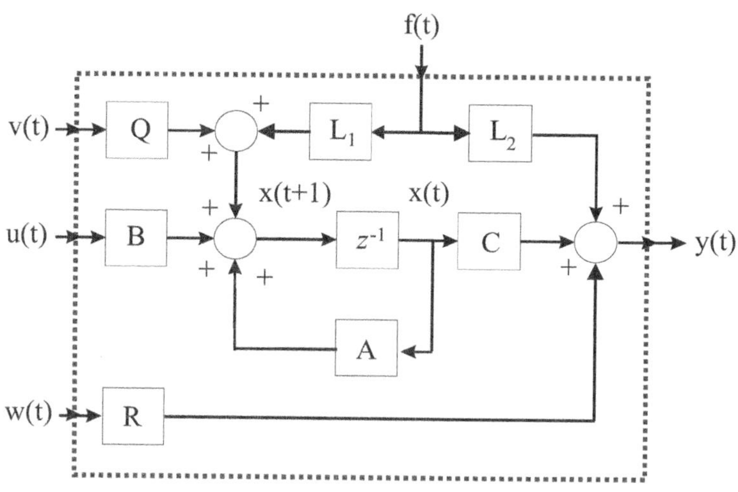

Figure 1.14. *MIMO process with faults and noises*

For the state estimation error, the following equations hold if the disturbances $v(t) = 0$ and $w(t) = 0$:

$$e_x(t+1) = (A - HC)e_x(t) + L_1 f(t) - HL_2 f(t) \qquad [1.38]$$

and the OE $e(t)$ becomes

$$e(t) = Ce_x(t) + L_2 f(t). \qquad [1.39]$$

The vector $f(t)$ represents *additive faults* because they influence $e(t)$ and $x(t)$ by a summation.

When sudden and permanent faults $f(t)$ occur, the state estimation error will deviate from zero. $e_x(t)$ and $e(t)$ show dynamic behaviors, which are different for $L_1 f(t)$ and $L_2 f(t)$. Both $e_x(t)$ or $e(t)$ can be taken as residuals.

In particular, the residual $e(t)$ is the basis for different fault detection methods based on output estimation.

For the generation of a residual with special properties, the design of the observer feedback matrix H is of interest (Chen and Patton 1999; Liu and Patton 1998).

Limiting conditions are the stability and the sensitivity against disturbances $v(t)$ and $w(t)$. If the signals are affected by noise, the Kalman filter must be used instead of classical observers (Jazwinski 1970).

If faults appear as changes ΔA or ΔB of the parameters, the process behavior becomes:

$$\begin{cases} x(t+1) = (A + \Delta A)x(t) + (B + \Delta B)u(t) \\ y(t) \qquad\quad = Cx(t) \end{cases} \qquad [1.40]$$

while the state $e_x(t)$ and the output estimation $e(t)$ errors are given as:

$$\begin{cases} e_x(t+1) = (A - HC)e_x(t) + \Delta A x(t) + \Delta B u(t) \\ e(t) \qquad\quad\;\; = Ce_x(t). \end{cases} \qquad [1.41]$$

The changes ΔA and ΔB are then *multiplicative faults* (Isermann 1997; Patton *et al.* 2000).

In this case, the changes in the residuals depend on the parameter changes, as well as input and state variable changes. Hence, the influence of parameter changes on the residuals is not as straightforward as in the case of the additive faults $f(t)$.

The following observer-based fault detection schemes and configurations are briefly summarized and recalled (Isermann 1997; Willsky 1976; Patton *et al.* 1989, 2000; Chen and Patton 1999).

1) **Dedicated observers for MIMO processes**

- *Observer excited by one output*: one observer is driven by one sensor output. The other outputs $\hat{y}(t)$ are reconstructed and compared with measured outputs $y(t)$. This allows the detection of single output sensor faults (Clark 1978).

- *Bank of observers, excited by all outputs*: several observers are designed for a definite fault signal and are detected by a hypothesis test (Willsky 1976).

- *Bank of observers, excited by single outputs*: several observers for single sensors outputs are used. The estimated outputs $\hat{y}(t)$ are compared with the measured outputs $y(t)$. this allows the detection of multiple sensor faults (Dedicated Observer Scheme) (Clark 1978).

- *Bank of observers, excited by all outputs except one*: as previously mentioned, but each observer is excited by all outputs except one sensor output, which is supervised (Generalized Observer Scheme) (Wünnenberg and Frank 1987; Frank 1993).

2) **Fault detection filters for MIMO processes**

- The feedback H of the state observer in equation [1.34] is chosen so that particular fault signals $L_1 f(t)$ change in a definite direction and fault signals $L_2 f(t)$ change in a definite plane (Beard 1971; Jones 1973; Chen and Speyer 1999).

With directional residual vectors, the fault isolation problem consists of determining which of the known fault signature directions the residual vector lies the closest to. The original form of the "failure detection filter" was proposed by Beard (1971) and Jones (1973) to generate directional residual vectors. Many more straightforward methods have followed, including methods to achieve a "robust fault detection filter" (Chen *et al.* 1996b).

The fault (or failure) detection is a class of Luenberger observers with a specially designed feedback gain matrix. It allows output estimation errors with directional characteristics associated with some known fault directions, to be obtained.

These fault detection methods mostly require several measurable output signals and make use of the internal analytical redundancy of multivariable systems. Recently, it was proposed to improve their robustness with respect to process parameter changes and unknown input signals $v(t)$ and $w(t)$ (Patton and Chen 1994; Chen *et al.* 1996b; Chung and Speyer 1998; Chen and Speyer 1999).

This can be reached, for example, through filtering the OE of the observer by

$$r(t) = We(t) \qquad [1.42]$$

together with a special design of the observer feedback matrix H.

3) **Output observers**

Another possibility is the use of output observers, or the so-called generalized observers, for example, the unknown input observers (UIO), in the reconstruction of the output signals, if the estimate of the state variable $\hat{x}(t)$ is not of primary interest.

In this context, it is worth mentioning the paper by Chen *et al.* (1996b) concerning the design of output observers for robust FDI using the eigenstructure assignment method.

Through a linear transformation

$$z(t) = Tx(t) \qquad [1.43]$$

the state–space representation of the observer becomes

$$\hat{z}(t+1) = F\hat{z}(t) + Ju(t) + Gy(t) \qquad [1.44]$$

and the residual is determined by

$$r(t) = W_z\hat{z}(t) + W_y y(t). \qquad [1.45]$$

This situation is depicted in Figure 1.15.

The state estimation error

$$e_x(t) = \hat{z}(t) - z(t) = \hat{z}(t) - Tx(t) \qquad [1.46]$$

and the residuals $r(t)$ are then designed, such that they are independent of the process states $x(t)$, the known input $u(t)$ and the unknown inputs $v(t)$ and $w(t)$, as depicted in Figure 1.14.

In this way, the residuals are dependent only on fault signals $f(t)$ (Chen and Patton 1999; Patton and Chen 1994; Chen *et al.* 1996b; Gertler 1998; Patton *et al.* 2000).

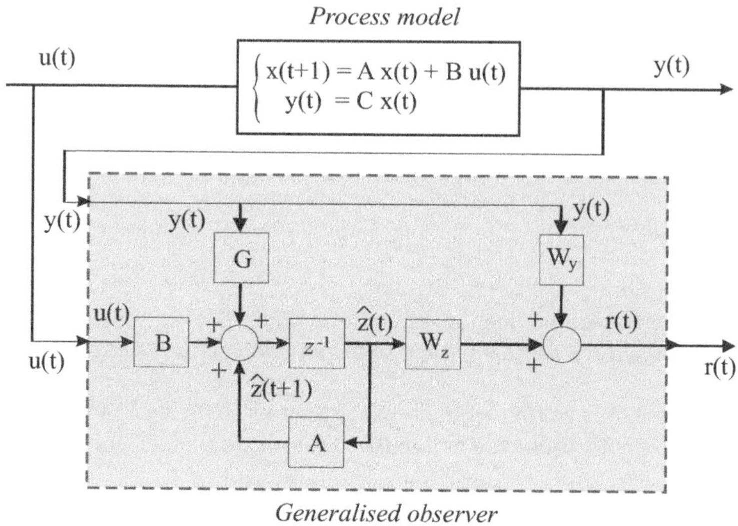

Figure 1.15. *Process and output observer*

1.5.3. *Fault detection via parity equations*

The basic idea of the parity relations approach is to provide a proper check of the parity (consistency) of the measurements acquired from the monitored system.

In the early development of fault diagnosis, the parity vector (relation) approach was applied to static or parallel redundancy schemes (Potter and Suman 1977), which may be obtained directly from measurements (hardware redundancy) or from analytical relations (analytical redundancy). A survey of these methods can be found in Ray and Luck (1991).

In the case of hardware redundancy, two methods can be exploited to obtain redundant relations. The first requires the use of several sensors having identical or similar functions to measure the same variable. The second approach consists of dissimilar sensors to measure different variables, but with their outputs being relative to each other.

Even if these techniques have been successfully applied for fault diagnosis (Potter and Suman 1977; Daly *et al.* 1979), this section is focused on analytical forms of redundancy.

A straightforward model-based method of fault detection is to take a model $G_M(z) = \frac{\hat{A}(z)}{\hat{B}(z)}$ and to run it in parallel to the process described by $G_P(z) = \frac{A(z)}{B(z)}$, thereby forming an error vector $r(z)$

$$r(z) = \left(\frac{A(z)}{B(z)} - \frac{\hat{A}(z)}{\hat{B}(z)}\right)u(z). \qquad [1.47]$$

The methodology described here is depicted in Figure 1.16.

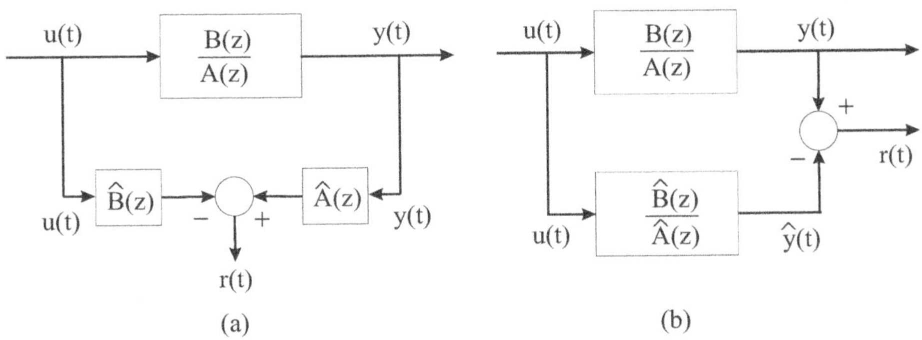

Figure 1.16. *Parity equation methods: (a) EE and (b) OE*

However, as for observers, the model parameters and structure of the monitored process have to be known *a priori*.

With reference to Figure 1.5, if

$$G_M(z) = G_P(z) \text{ i.e. } \frac{\hat{A}(z)}{\hat{B}(z)} = \frac{A(z)}{B(z)} \qquad [1.48]$$

for additive input $f_u(z)$ and output $f_y(z)$ faults, the $r(z)$ error then becomes

$$r(z) = \frac{A(z)}{B(z)} f_u(z) + f_y(z). \qquad [1.49]$$

According to Figure 1.16, another possibility is to generate a polynomial error:

$$\begin{aligned} r(z) &= \hat{A}(z)y(z) - \hat{B}(z)u(z) \\ &= B(z)f_u(z) + A(z)f_y(z). \end{aligned} \qquad [1.50]$$

In both cases, different time responses are obtained for an additive input or output fault.

Moreover, the error vector $r(z)$ computed by equation [1.49] corresponds to the OE of the parameter estimation method computed by equation [1.28].

On the other hand, $r(z)$ in equation [1.50] concerns the equation error of equation [1.21].

Equations [1.49] and [1.50] generate residuals and are called *parity equations* (Gertler 1991) under the assumptions of fault occurrence and of exact agreement between process and model.

However, within the parity equations, the model parameters are assumed to be known and constant, whereas the parameter estimations can vary the parameters of $\hat{A}(z)$ and $\hat{B}(z)$ in order to minimize the residuals.

Moreover, for the generation of specific characteristics of the parity vector $r(z)$ and for obtaining fault detection and isolation properties, the residuals can be filtered according to matrix $G_f(z)$ to compute the vector $r_f(z)$ (Gertler 1991; Patton and Chen 1994; Patton *et al.* 2000):

$$r_f(z) = G_f(z)r(z). \qquad [1.51]$$

Equations [1.49]–[1.51] can therefore be used to implement and design the residual generation system, in order to meet fault detection and isolation specifications (Gertler 1998).

However, only one residual can be generated for SISO processes and it is therefore not easy to distinguish between different faults.

On the other hand, more freedom in the design of parity equations can be obtained when intermediate signals can be measured for SISO processes, as shown in Figure 1.5, or for MIMO systems.

As an extension of the parity equation method, the parity relation concept presented here can be generalized (Chow and Willsky 1984; Lou *et al.* 1986; Patton and Chen 1994) and then extended to state–space descriptions, as shown in Gertler (1998) for discrete-time models.

The redundancy relations are now specified mathematically as follows.

Given the system

$$\begin{cases} x(t+1) = Ax(t) + Bu(t) \\ y(t) \quad\;\; = Cx(t) \end{cases} \qquad [1.52]$$

by substituting the second of equations [1.52] into the first one and delaying several times, the following system is obtained:

$$\begin{bmatrix} y(t) \\ y(t+1) \\ y(t+2) \\ \vdots \end{bmatrix} = \begin{bmatrix} C \\ CA \\ CA^2 \\ \vdots \end{bmatrix} x(t) + \begin{bmatrix} 0 & 0 & 0 \ldots \\ CB & 0 & 0 \ldots \\ CAB & CB & 0 \ldots \\ \vdots & \vdots & \vdots \ddots \end{bmatrix} \begin{bmatrix} u(t) \\ u(t+1) \\ u(t+2) \\ \vdots \end{bmatrix} \quad [1.53]$$

$$Y_f(t) = Tx(t) + QU_f(t). \quad [1.54]$$

In order to remove the non-measurable states $x(t)$ and obtain a parity vector useful for FDI, equation [1.53] is multiplied by W, such that

$$WT = 0. \quad [1.55]$$

This leads to residuals

$$r(t) = WY_f - WQU_f(t) \quad [1.56]$$

as shown in Figure 1.17.

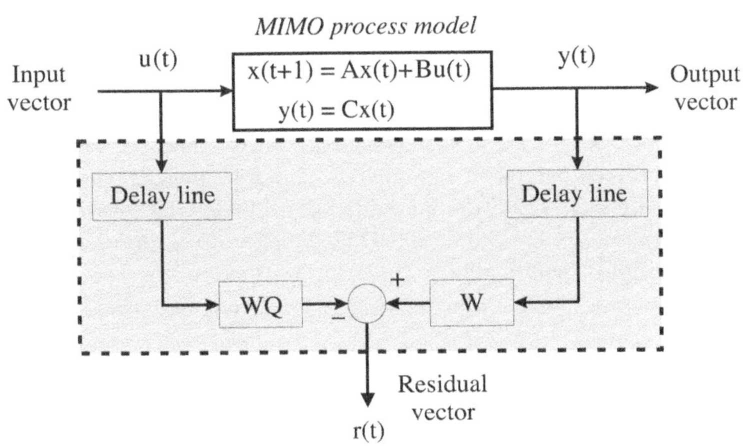

Figure 1.17. *Parity EE methods for a MIMO model*

The filtered input and output vectors U_f and Y_f are obtained by delaying the corresponding signals.

The design of the matrix W gives some freedom to generate a structured set of residuals.

One possibility is to select the elements of W such that one measured variable has no impact on a specific residual. Then, this residual remains small in the case of an additive fault on this variable, and the other residuals increase (Patton and Chen 1994; Chen and Patton 1999).

Finally, because of the previous results, it is clear that some correspondence exists between the parity relation and observer-based methods. This aspect was first pointed out by Massoumnia (1986) and was later demonstrated by Wünnenberg (1990) and Patton *et al.* (1989).

The problem was re-examined in detail by Patton and Chen (1994), and the equivalence under different conditions and in different meanings was discussed. It was shown that the parity relation approach is equivalent to the use of a dead-beat observer.

This implies that the parity relation scheme provides less design flexibility when compared with methods, which are based on observers without any restriction.

More recently, a comparison between observer-based and parity space techniques was proposed (Delmaire *et al.* 1999). Both of the methods were first explored for SISO systems and therefore extended the comparison to MIMO systems. The comparison was performed using linear discrete-time models.

In particular, considering the MIMO systems described by the estimated input–output discrete-time forms (e.g. ARX or Auto Regressive Moving Average eXogenous [ARMAX] models) of equations [1.49] and [1.50], leads to a representation in which parameter redundancy cannot be avoided. To overcome this drawback, Delmaire *et al.* (1999) proposed to use observers designed from identified canonical state-space forms (Guidorzi 1975). Moreover, in the case of parameter redundancy, multiple identifications of some parameters may occur, leading to inconsistent estimations that might produce inconsistent FDI decisions (Delmaire *et al.* 1999).

This states again the FDI capabilities of the observer-based methods with respect to parity relation schemes.

1.6. Change detection and symptom evaluation

When the residual generation stage has been performed, the second step requires the examination of symptoms in order to determine if any faults have occurred.

As shown in equation [1.17], a decision process may consist of a simple threshold test on the instantaneous values of the moving averages of residuals.

On the other hand, because of the presence of noise, disturbances and other unknown signals acting upon the monitored system, the decision-making process can exploit statistical methods.

In this case, the measured or estimated quantities, such as signals, parameters, state variables or residuals, are usually represented by stochastic variables $r(t) = \{r_i(t)\}_i^q$, with mean value and variance (Willsky 1976)

$$\bar{r}_i = E\{r_i(t)\}; \qquad \bar{\sigma}_i^2 = E\{[r_i(t) - \bar{r}_i]^2\} \qquad [1.57]$$

as normal values for the fault-free process.

Analytic symptoms are then obtained as changes:

$$\Delta r_i = E\{r_i(t) - \bar{r}_i\}; \qquad \Delta\sigma_i = E\{\sigma_i(t) - \bar{\sigma}_i\} \qquad [1.58]$$

with reference to the normal values. Usually, the time instant $t > t_f$ represents the unknown instant of the fault occurrence.

In order to separate normal behavior from faulty behavior, a fixed threshold Δr_{tol} usually defined as

$$\Delta r_{tol} = \epsilon\bar{\sigma}_r, \qquad \epsilon \geq 2 \qquad [1.59]$$

has to be selected.

By a proper choice of ϵ, a compromise has to be made between the detection of small faults and false alarms.

Another class of methods can be exploited for detecting residual changes due to faults. Therefore, techniques of change detection, such as a likelihood-ratio test or Bayes decision, a run–sum test, are commonly used (Isermann 1984; Basseville and Benveniste 1986; Basseville and Nikiforov 1993).

Moreover, fuzzy or adaptive thresholds may improve the binary decision (Chen and Patton 1999; Patton *et al.* 2000).

Finally, when several variables change, classification methods are used. In a multidimensional space, the symptom vector

$$\Delta r = [\Delta r_1 \; \Delta r_2 \; \cdots \; \Delta r_q] \qquad [1.60]$$

belongs to a q-dimensional space and its direction depends on the fault occurrence.

In this case, the process of residual evaluation consists of determining the direction as well as the distance of Δr from the origin. Geometrical distance methods (Carpenter and Grossberg 1987; Tou and Gonzalez 1974) or artificial neural networks (NNs) (Himmelblau *et al.* 1991; Meneganti *et al.* 1998) can hence be applied.

The generation and evaluation of analytic symptoms concludes the task of fault detection within the framework of model-based fault diagnosis, as shown in Figure 1.8.

1.7. Residual generation robustness problem

Although the analytical redundancy method for residual generation has been recognized as an effective technique for detecting and isolating faults, the critical problem of unavoidable modeling uncertainty has not been fully solved.

The main problem regarding the reliability of FDI schemes is the modeling uncertainty, which is due, for example, to process noise, parameter variations and nonlinearities.

On the other hand, all model-based methods use a model of the monitored system to produce the symptom generator. If the system is not complex and can be described accurately by the mathematical model, FDI is directly performed by using a simple geometrical analysis of residuals. In real industrial systems, however, the modeling uncertainty is unavoidable.

The design of an effective and reliable FDI scheme for residual generation should take into account the modeling uncertainty with respect to the sensitivity of the faults. Therefore, the task of the design of an FDI system is thus to generate residuals that are *robust* (Chow and Willsky 1984; Ding and Frank 1990; Frank 1994b; Frank and Ding 1997; Patton and Chen 1994).

Several papers addressed this problem. For example, optimal robust parity relations were proposed (Chow and Willsky 1984; Chung and Speyer 1998; Chen and Speyer 1999; Lou *et al.* 1986) and the threshold selector concept was introduced (Emami-Naeini *et al.* 1988). Robust FDI using the disturbance decoupling technique was also used (Patton and Chen 1994; Chen *et al.* 1996b). Patton and Chen's approach is an interesting contrast to Chow and Willsky's method, which seems to minimize the modeling uncertainty over several points of operation. Patton and Chen deal directly with this problem by estimating the optimum unknown input distribution matrix over a range of operating points and exploiting the eigenstructure assignment approach (Patton and Chen 1994; Chen and Patton 1999).

The model-based FDI technique requires a high accuracy mathematical description of the monitored system. The better the model represents the dynamic behavior of the system, the better the FDI precision will be. If an FDI method can be developed, which is insensitive to modeling uncertainty, a very accurate model is not necessarily needed.

All uncertainties are summarized as disturbances acting on the system. Although the disturbance vector is unknown, its distribution matrix can be obtained by an identification procedure. Under this assumption, the "disturbance de-coupling" principle can be exploited to design a robust FDI scheme.

In order to summarize the approach to the robustness problem, the state–space model of the monitored system should be considered (Patton and Chen 1993):

$$\begin{cases} x(t+1) = (A + \Delta A)\, x(t) + (B + \Delta B)\, u(t) + E_1 \varepsilon(t) + R_1 f(t) \\ y(t) \quad\;\; = Cx(t) + E_2 \varepsilon(t) + R_2 f(t) \end{cases} \qquad [1.61]$$

where $\varepsilon(t)$ is the disturbance vector, and E_1 and E_2 are the known and unknown input distribution matrices. The matrices ΔA and ΔB are the parameter errors or variations that represent modeling errors.

The discrete transfer matrix description between the output $y(t)$ and input $u(t)$ of the system [1.61] is then

$$y(z) = (G_u(z) + \Delta G_u(z))\, u(z) + G_\varepsilon(z)\varepsilon(z) + G_f(z) f(z) \qquad [1.62]$$

where $\Delta G_u(z)$ is used to describe modeling errors, while both $\Delta G_u(z)$ and $G_\varepsilon(z)$ represent modeling uncertainty.

With reference to the residual generator in Figure 1.10 and described by equation [1.14], the z-domain residual vector is rewritten as

$$r(z) = H_y(z) G_f(z) f(z) + H_y(z) G_\varepsilon(z)\varepsilon(z) + H_y(z)\Delta G_u(z) u(z). \qquad [1.63]$$

With respect to equation [1.14], the terms $H_y(z)G_\varepsilon(z)$ and $H_y(z)\Delta G_u(z)$ cannot be deleted.

Both faults and modeling uncertainty (disturbance and modeling error) affect the residual, and hence discrimination between these two effects is difficult.

The principle of disturbance de-coupling for robust residual generation requires that the residual generator satisfies

$$H_y(z) G_\varepsilon(z) = 0 \qquad [1.64]$$

in order to achieve total de-coupling between residual $r(z)$ and disturbance $\varepsilon(z)$.

This property can be achieved by using the unknown input observer (Watanabe and Himmelblau 1982; Wünnenberg and Frank 1987; Chen *et al.* 1996b; Frank *et al.* 2000), optimal (robust) parity relations (Chow and Willsky 1984; Lou *et al.* 1986; Wünnenberg 1990; Wünnenberg and Frank 1990; Frank *et al.* 2000) or alternatively, the eigenstructure assignment approach (Patton *et al.* 1986; Patton and Chen 1991b, 2000; Liu and Patton 1998; Duan *et al.* 2002).

These approaches can be integrated with different system identification tools for achieving robust residual generation schemes. Hence, for disturbance de-coupling approaches in FDI, the aim is to completely eliminate the disturbance effects from the residual. However, the complete elimination of disturbance effects may not be possible due to the lack of degree of freedom. Moreover, it may be problematic, in some cases, because the fault effect may also be eliminated. Hence, an appropriate criterion for robust residual design should take into account the effects of both modeling error and faults. There is a tradeoff between sensitivity to faults and robustness to modeling uncertainty, and hence robust residual generation can be considered as a *multi-objective optimization problem* (Chen and Patton 1999, Chapter 6). It consists of the maximization of fault effects and the minimization of uncertainty effects (Wünnenberg 1990; Frank *et al.* 2000).

1.7.1. *FDI H_∞ approach*

Therefore, the approach to the design of optimal residuals requires the satisfaction of a set of objectives. These objectives are essential for achieving a robust diagnosis of incipient faults. If such joint optimization problems, which can be also expressed in the frequency domain, were reformulated for satisfying a set of inequalities on the performance indices, Genetic Algorithms (GA) (Goldberg 1989; Davis 1991) and Linear Matrix Inequalities (LMI) (Boyd *et al.* 1994) can be successfully exploited to search the optimal solution (Chen *et al.* 1996a, 1997; Hou and Patton 1997; Chen and Patton 1999, 2001).

Disturbance de-coupling can also be achieved using frequency domain design techniques. As an example, the robust fault detection problem can be managed by using the standard H_∞ filtering formulation (Ding and Frank 1990; Hou and Patton 1996; Frank and Ding 1997).

With this method, the minimization of the disturbance effect on the residual is formulated as a standard H_∞ filtering problem (Chen and Patton 2000; Frank *et al.* 2000). On the other hand, the so-called H_∞/H_- approach can also be exploited (Hou and Patton 1996, 1997; Frank *et al.* 2000; Chen and Patton 2000).

Among the many ways of eliminating or minimizing disturbance and modeling error effects on the residual, and hence achieving robustness in FDI (Patton *et al.*

2000), H_∞ optimization is a robust design method with the original motivation firmly rooted in the consideration of various uncertainties, especially the modeling errors. It is reasonable to seek an application of this technique in the robust design of FDI systems. Therefore, the H_∞ optimization method can be successfully exploited for a robust residual generation of FDI.

The early work of using H_∞ optimization techniques for robust FDI was based on the use of the factorization approach (Ding and Frank 1990; Ding *et al.* 2000). The factorization-based H_∞ optimization technique is useful in solving FDI problems. However, the more elegant and advanced H_∞ optimization methods are based on the use of the Algebraic Riccati Equation (ARE) (Zhou *et al.* 1996). Mangoubi *et al.* (1992) first solved the robust FDI estimation problem using the ARE approach, via the use of the H_∞ and μ robust estimator synthesis methods developed by Appleby *et al.* (1991). A direct formulation of the FDI problem as a robust H_∞ filter design problem, with the solution of an ARE was given in Edelmayer *et al.* (1997). To deal with modeling errors as well as disturbances in robust FDI design, Niemann and Stoustrup (1996) introduced modeling error blocks into the standard H_∞ observer design. The weighting factors are then introduced in the problem formulation for finding an optimal FDI solution. This is further extended to nonlinear systems, where the nonlinearity is treated in the same way as a modeling error block (Stoustrup and Niemann 1998; Stoustrup *et al.* 1997).

The majority of studies discussed so far involve the use of a slightly modified H_∞ filter for the residual generation, that is, the design objective is to minimize the effect of disturbances and modeling errors on the estimation error and subsequently on the residual. However, robust residual generation is different from the robust estimation because it does not only require the disturbance attenuation. The residual has to remain sensitive to faults while the effect of disturbance is minimized. Sauter *et al.* (1997) studied this problem where the fault sensitivity is enhanced by applying an optimal post-filter to the "primary residual". The problem of enhancing fault sensitivity while increasing robustness against disturbances and modeling errors was studied extensively by Sadrnia *et al.* (1997). The essential idea is to reach an acceptable compromise between disturbance robustness and fault sensitivity. In the beginning, an observer with very small disturbance sensitivity bound is designed via an ARE. Then, the fault sensitivity is checked. If the fault sensitivity is too small, the disturbance robustness requirement should be relaxed, that is, to design another optimal observer with an increased disturbance sensitivity bound. This procedure is likely to be repeated several times. The final goal is to find a design that provides the maximum ratio between fault sensitivity and disturbance sensitivity.

Recently, Chen and Patton (1999, 2000) have formulated the robust residual generation problem within the standard H_∞ filtering framework, that is, to generate the residual whose sensitivity to disturbances is minimized. To facilitate reliable FDI, the residual sensitivity to faults has to be maintained (or maximized), in addition to the minimization of the disturbance sensitivity. This problem was solved via the minimization of the difference between the residual and the fault against the disturbance and the fault, that is, the objective is to replicate the fault using the residual. In this way, the residual sensitivity to the fault is indirectly maximized. The residual sensitivity to the modeling error can be minimized if the modeling error is approximately represented by the disturbance vector with the estimated distribution matrix (Chen and Patton 1999). However, the modeling error can be handled directly using standard H_∞. In Chen and Patton (1999, 2000), the method of including the modeling error in the robust residual design within the standard H_∞ framework was shown.

In case the condition of equation [1.64] is not satisfied, only an approximate de-coupling can be obtained. In particular, the performance index (Ding and Frank 1991)

$$J_d = \frac{\| H_y(w)G_\varepsilon(w) \|}{\| H_y(w)G_f(w) \|} \qquad [1.65]$$

can be defined by considering the w-transformation in the frequency domain for discrete-data systems (Kuo 1995).

Generally speaking, the robust FDI approach can be approached in different ways. It is therefore important to mention the design principle of residual generators under a certain performance index (Basseville 1997; Frank *et al.* 2000). This is indeed a reasonable extension of the unknown input residual generator design, in which, instead of full de-coupling, a compromise between the robustness and sensitivity is made.

It is worth focusing the attention to this scheme, due to its important role in theoretical studies and its relationship to the residual evaluation and integrated design of FDI systems. Since the goal of residual generation is to enhance the robustness of the residual to the model uncertainty, without loss of its sensitivity to the faults, the minimization of performance index (Frank *et al.* 2000)

$$J = \frac{\|\frac{\partial r}{\partial d}\|}{\|\frac{\partial r}{\partial f}\|} \text{ or } J = \|\frac{\partial r}{\partial d}\| \text{ with } \|\frac{\partial r}{\partial f}\| > \alpha \qquad [1.66]$$

is widely recognized as a suitable design objective. Associated with the norm used, the type of the residual generator and the mathematical tool adopted, a number of optimization approaches have been developed (Frank *et al.* 2000). Recently, Ding *et al.* (2000) derived a unified solution for a number of optimization problems, and thus provided a satisfactory solution to the above-defined optimization problem 10 years after it was first proposed. In Frank *et al.* (2000), a brief review of the state of art of the solutions can be found, whereas in Hou and Patton (1996, 1997) and Frank *et al.* (2000), the H_∞/H_- method is detailed.

By minimizing the performance index of [1.66] over a specified range in the w-transformed plane (Kuo 1995), an approximate de-coupling design can be achieved (Ding and Frank 1990; Patton and Hou 1997; Frank and Ding 1997; Ding *et al.* 1999). Moreover, the approximated design for optimal disturbance de-coupling can also be solved in the time domain (Wünnenberg 1990; Chen *et al.* 1993).

According to the norm selected, by minimizing the performance index of [1.66] over a specified range, an approximate de-coupling design can be achieved (Ding and Frank 1990; Patton and Hou 1997; Frank and Ding 1997; Ding *et al.* 1999).

1.7.2. *Active and passive disturbance de-coupling*

Regarding this issue, the approximated design for optimal disturbance de-coupling can also be solved in the time domain (Wünnenberg 1990; Chen *et al.* 1993).

On the other hand, with reference to the modeling errors in equation [1.63], represented by the term $\Delta G_u(z)$, the robust problem is more difficult to solve.

Two main techniques have been described by Patton and Chen. In the first case, the uncertainty is taken into account at the residual design stage (Chen *et al.* 1996b); this is known as *active robustness* in fault diagnosis (Patton and Chen 1994).

One attempts to take into account the uncertainty in the design of the residual (see, e.g. Chen *et al.* 1993; Frank 1996); it is known as *active robustness*.

The active way of achieving a robust solution is to approximate uncertainties, that is, by approximately representing modeling errors as disturbances (Chen and Patton 1999):

$$\Delta G_u(z)u(z) \approx G_d(z)d(z) \qquad [1.67]$$

where $d(z)$ is an unknown vector and $G_d(z)$ is an estimated transfer function. When this approximate structure is exploited to design disturbance de-coupling residual generators, robust FDI can be achieved.

The second approach called *passive robustness* makes use of a residual evaluator with an adaptive threshold. As a simple example, it is assumed that the residual generation uncertainty [1.63] is only represented by modeling errors.

The fault-free residual $r(z)$ is given as

$$r(z) = H_y(z)\Delta G_u(z)u(z). \qquad [1.68]$$

Under the assumption that the modeling errors are bounded by a value δ, such that

$$\| \Delta G_u(w) \| \leq \delta \qquad [1.69]$$

an adaptive threshold $\varepsilon(t)$ can be generated by a linear system

$$\varepsilon(t) = \delta H_y(z) u(z) \qquad [1.70]$$

In such a case, the threshold $\varepsilon(t)$ is no longer fixed, but depends on the input $u(t)$, thus being adaptive to the system operating point. A fault is then detected if

$$\| r(t) \| > \| \varepsilon(t) \| \qquad [1.71]$$

A robust FDI technique with a threshold adaptor or selector is therefore briefly recalled (Clark 1989; Emami-Naeini *et al.* 1988; Ding and Frank 1991). This method represents a passive approach since no effort is made to design a robust residual.

Even if disturbance de-coupling methods for robust FDI have been studied extensively, their effectiveness regarding real problems has not been fully demonstrated.

The main difficulty arises as most of the disturbances only account for a small percentage of the uncertainty in the real system. The presented disturbance decoupling methods cannot be directly applied to the systems with other uncertainties, such as modeling errors.

The estimation and approximate representation of modeling errors, as well as other uncertain factors, as the disturbance term provides a practical way to tackle the robustness issue for real plants.

This chapter provides some considerations on different approaches for representing modeling errors and other uncertain factors via the disturbance term with an estimated distribution matrix. As addressed in a system identification framework, this identified distribution matrix will be used for the design of the disturbance de-coupled residual, in order to solve the robust FDI problem.

1.8. Fault diagnosis technique integration

Several FDI techniques have been developed and their application shows different properties with respect to the diagnosis of different faults in a process. In order to achieve a reliable FDI technique, a good solution consists of a proper integration of several methods, which take advantage of the different procedures (Isermann 1994; Isermann and Ballé 1997).

Furthermore, a comprehensive approach to fault diagnosis should exploit a knowledge-based treatment of all available analytical and heuristic information. This successful approach can be performed by an integrated method to knowledge-based fault diagnosis.

1.8.1. *Fuzzy logic for residual generation*

As stated in section 1.2, model-based FDI consists of two stages, residual generation and decision-making.

The first block is exploited to generate residuals by means of the available inputs and outputs from the monitored system.

For the first step, classical fault diagnosis model-based methods can exploit state–space of input–output dynamic models of the process under investigation. Within this framework, faults are supposed to appear as changes on the system state or output caused by malfunctions of the components, as well as of the sensors. Such fault indices are often monitored using estimation techniques.

The main problem with these techniques is that the precision of the process model affects the accuracy of the detection and isolation system, as well as the diagnostic sensibility.

On the other hand, the majority of real industrial processes are nonlinear (Chen and Patton 1999; Gertler 1998; Patton and Chen 1997) and cannot be modeled by using a single model for all operating conditions.

Since a mathematical model is a description of system behavior, accurate modeling for a complex nonlinear system is very difficult to achieve in practice. Sometimes, for some nonlinear systems, it can be impossible to describe them by analytical equations. Moreover, sometimes the system structure or parameters are not precisely known and if diagnosis has to be based primarily on heuristic information, no qualitative model can be set up.

Because of these assumptions, fuzzy system theory seems to be a natural tool to handle complicated and uncertain conditions (Babuška 1998).

Instead of exploiting complicated nonlinear models obtained by modeling techniques, it is also possible to describe the plant by a collection of local affine fuzzy and non-fuzzy models (Leontaritis and Billings 1985a, 1985b; Takagi and Sugeno 1985), whose parameters are obtained by identification procedures.

The second stage of model-based FDI consists of a logic decision process that transforms residual signal information (quantitative knowledge) into qualitative statements (faulty or normal working conditions). Therefore, the problem of decision making can be treated in a novel way by means of fuzzy logic.

As noise contamination and uncertainty affect the residuals, even in fault-free conditions, they fluctuate and become unequal to zero. This common situation, which may hide the fault effects, can be handled by means of the fuzzy logic framework.

The interesting feature of fuzzy logic is that it represents a powerful tool for describing vague and imprecise facts, and is therefore well suited for applications where complete information about fault and system is not available to the designer.

Even if much effort has been spent on trying to decrease the uncertainty associated with quantitative residual generation, it is impossible to fully eliminate the effect of uncertainty. On the basis of this limitation, the residual evaluation problem consists of making the correct decision with respect to uncertain information. Fuzzy logic can be a suitable tool for this task. For instance, a lot of processes can be managed by humans heuristically, since an analytical description is impossible to use. Fuzzy logic can express expert knowledge in the form of a rule-based knowledge format.

The introduction of fuzzy logic can improve the decision making in order to provide reliable FDI methods, which are applicable for real industrial systems.

As an example, fuzzy logic can be exploited for residual evaluation, mainly in the decision-making stage for releasing the final yes–no decision (Ulieru and Isermann 1993; Frank 1994a; Meneganti *et al.* 1998).

Rule-based expert systems have, therefore, been investigated very intensively for fault detection and diagnosis problems (Rich and Venkatasubramanian 1987; Kramer 1987; Patton *et al.* 1989, 2000). Fault diagnosis using rule-based systems needs a database of rules, and the accuracy of diagnosis depends on the rules. Moreover, creating a rich and detailed database of rules is usually a time-consuming task and many process experts are needed.

It should finally be pointed out how the fuzzy approach in FDI can solve the problem at two levels: first, fuzzy descriptions are used to generate symptoms and then, the fault detection and isolation is achieved again using fuzzy logic (Dexter and Benouarets 1997; Isermann 1998).

1.8.2. *Neural networks for fault diagnosis*

Quantitative model-based fault diagnosis generates symptoms on the basis of the analytical knowledge of the process under investigation. In most cases however, this does not provide enough information to perform an efficient FDI, that is, to indicate the location and the mode of the fault.

A typical integrated fault diagnosis system uses both analytical and heuristic knowledge of the monitored system. The knowledge can be processed in terms of residual generation (analytical knowledge) and feature extraction (heuristic knowledge). The processed knowledge is then provided to an inference mechanism, which can comprise residual evaluation, symptom observation and *pattern recognition*.

In particular, when the process model is only known to a certain extent of precision, the pattern recognition method can provide a convenient approach to solve the fault identification problem, that is, to determine the size of the fault (Himmelblau 1978; Pau 1981).

In recent years, NNs have been used successfully in pattern recognition, as well as system identification, and they have been also been proposed as a possible technique for fault diagnosis.

NN can handle nonlinear behavior and partially known processes because they learn the diagnostic requirements by means of the information of the training data.

NNs are noise tolerant and their ability to generalize the knowledge, as well as adapt during use are extremely interesting properties (Hoskins and Himmelblau 1988; Dietz *et al.* 1989; Venkatasubramanian and Chan 1989; McDuff and Simpson 1990; Chen *et al.* 1990). Some example processes were considered, in which FDI was performed by an NN using input and output measurements. In these works, the NN is trained to identify the fault from measurement patterns, however the classification of the individual measurement pattern is not always unique in dynamic situations, therefore the straightforward use of NN in fault diagnosis of dynamic plants is not practical and other approaches should be investigated.

An NN could be exploited in order to find a dynamic model of the monitored system or connections from faults to residuals. In the latter case, the NN is used as a pattern classifier or nonlinear function approximator. In fact, artificial NNs are capable of approximating a large class of functions for the fault diagnosis of an industrial plant.

Under these considerations, this book also considers the identification of fuzzy and non-fuzzy models for the system under diagnosis, as well as the application of NN as function approximators.

Quantitative and qualitative approaches have a lot of complementary characteristics, which can be suitably combined together to exploit their advantages and increase the robustness of quantitative techniques. The suggested combination can also minimize the disadvantages of the two procedures; in particular, it is important that partial knowledge deriving from qualitative reasoning is reduced by quantitative methods. Hence, the main aim of further research on model-based fault diagnosis is to find a way to properly combine these two approaches together to provide highly reliable diagnostic information.

1.8.3. *Neuro-fuzzy approaches to FDI*

The identification of multivariable processes can be interpreted as a problem of approximation to an input–output mapping. The mathematical model used in traditional methods is sensitive to modeling errors, parameter variation, noise and disturbance (Chen and Patton 1999; Patton *et al.* 2000). Process modeling has limitations, especially when the system is complex and uncertain, and the data are ambiguous and not rich in information.

As previously stated, NNs are known to approximate any nonlinear even dynamic function, given suitable weighting factors and architecture. Moreover, online training makes it possible to change the FDI system easily in cases where changes are made in the physical process or the control system. NNs can generalize when presented with inputs not appearing in the training data and make intelligent decisions in cases of noisy or corrupted data. They are also readily applicable to multivariable systems and have a highly parallel structure, which is expected to achieve a higher degree of fault tolerance. An NN can operate simultaneously on qualitative and quantitative data. NN can be very useful when no mathematical model of the system is available, that is, analytical models cannot be applied.

Almost all of the physical processes are dynamic in nature. Combining dynamic elements, such as filters and delays, yields a powerful modeling technique. But the NN operates as a "black box" with no qualitative/quantitative information available about the model it represents. Usually, engineers and operators want to visualize how the system is working and what rules govern its operation. There is also ambiguity about the performance of the NN in case of unexpected situations (Korbicz *et al.* 1999).

Fuzzy logic systems, on the other hand, have the ability to mimic the sensing, generalizing, processing, operating and learning abilities of a human operator. They offer a linguistic model of the system dynamics, which can be easily understood by certain rules. They also have inherent abilities to deal with imprecise or noisy data.

Fuzzy logic can be used with NNs (Chiang *et al.* 2001). A fuzzy neuron has the same basic structure as the artificial neuron, except that some or all of its components and parameters may be described through fuzzy logic. A fuzzy NN is built on fuzzy neurons or on standard neurons dealing with fuzzy data. A fuzzy NN is a connectionist model for the implementation and inference of fuzzy rules. There are many different ways to fuzzify an artificial neuron, which results in a variety of fuzzy neurons and fuzzy networks (Chiang *et al.* 2001; Nelles 2001).

Different neuro-fuzzy (NF) structures can therefore be designed to combine the advantages of both NNs and fuzzy logic (Patton *et al.* 1999; Calado *et al.* 2001). These structures have been successfully applied to a wide range of applications from industrial processes to financial systems, because of the ease of rule base design, linguistic modeling, application to complex and uncertain systems, inherent nonlinear nature, learning abilities, parallel processing and fault-tolerance abilities (Wu and Harris 1996; Ayoubi 1995). However, successful implementation depends heavily on prior knowledge of the system and the training data. There are three common methods of combining NNs with fuzzy logic:

1) fuzzification of the inputs or outputs of the NNs;

2) fuzzification of the interconnections of conventional NNs;

3) using NNs in fuzzy models where neurons provide the necessary membership functions and rule base.

Research effort has also been focused also on NF networks making use of B-spline functions, which can be used to identify the process using NN architecture and at the same time extract some qualitative knowledge of the system being modeled. Applications of the B-spline NNs show how to successfully integrate the approximation techniques of NNs and the qualitative approach of fuzzy logic. The operator can also include any heuristic knowledge about the plant. Unlike many other NF approaches, B-spline networks offer a simple and easy to build framework (Uppal and Patton 2000).

Once the NF network structure and parameters are identified, such network can be useful in general FDI schemes, which consists of residual generation and decision making (classification). In residual generation, the residual vector is determined in order to characterize each fault. The residual vector is then processed to determine the location and occurrence time of the faults, which is called decision making. Ideally, the models identify all classes of system behavior (Uppal and Patton 2000).

Therefore, the NF network is used to classify process faults that are assumed to be known *a priori*. Their corresponding data are available to the designer. The network then has as many outputs as classes of behavior. Hence, for a system with M classes of faults, the output of the network will be a vector of dimension $M+1$; this includes the models associated with the M faults, as well as the one corresponding to the healthy case.

To train the network, it is necessary to decompose it into a set of $(M+1)$ MIMO sub-models, if M is the number of faulty classes, and find the set of optimal weighting coefficients with each sub-model. When the network is used to classify a test point

$\{u(t), y(t)\}$, the network's output is a real vector of dimension $(M + 1)$. Under these assumptions, each component of that output vector (from 1 to M) is identified with a class of behavior, which can represent either the ith fault ($i = 1, \ldots, M$) or the nominal model. When the system is operating in its nominal condition, all of the network outputs are zero except the last one. However, when a specific fault develops in the system, the corresponding output will deviate from zero, whereas the output $i + 1$ becomes zero, confirming that the system is no longer healthy.

1.8.4. *Fault detectability and isolability*

A successful detection of a fault is obtained if a residual has the maximal sensitivity to its occurrence. Fault *detectability* conditions were stated in Chen and Patton (1999). Such a stage is followed by the fault isolation procedure, which allows us to distinguish a particular fault from others. While a single residual signal is sufficient to detect faults, a vector of residuals is usually required for fault isolation. Faults are distinguishable or *isolable* using the residual set if each residual is sensitive to a subset of faults, while remaining insensitive to the remaining faults. The design technique to obtain the so-called *structured residual set* will be shown briefly in the following sections.

All of the NF modeling structures combine, in a single framework, both numerical and symbolic knowledge about the process. Automatic linguistic rule extraction is a useful aspect of NF, especially when little or no prior knowledge about the process is available (Brown and Harris 1994; Jang and Sur 1995). For example, an NF model of a nonlinear dynamical system can be identified from the empirical data. This modeling approach can give us some insight about the nonlinearity and dynamical properties of the system.

The most common NF systems are based on two types of fuzzy models, TSK (Takagi and Sugeno 1985; Sugeno and Kang 1988; Mamdani 1976; Mamdani and Assilian 1995) combined with NN learning algorithms. TSK models use local linear models in the consequents, which are easier to interpret and can be used for control and fault diagnosis (Füssel *et al.* 1997; Isermann and Ballé 1997). Mamdani models use fuzzy sets or rules as consequents and therefore give a more qualitative description. The B-spline NN (with triangular basis functions) is the simplest of all of the Mamdani NF structures, but the large consequent rule set means that the method is not easy to use due to low transparency.

Many NF structures have been successfully applied to a wide range of applications, from industrial processes to financial systems, because of the ease of rule base design, linguistic modeling, application to complex and uncertain systems, inherent nonlinear nature, learning abilities, parallel processing and fault-tolerance abilities. However, successful implementation depends heavily on prior knowledge of the system and the empirical data (Ayoubi 1995).

NF networks by their intrinsic nature can handle a limited number of inputs and can usually be identified in a not very transparent way from the empirical data. Here, transparency corresponds to a more meaningful description of the process, that is, less rules with appropriate membership functions. In the Adaptive Neuro-Fuzzy Inference System (ANFIS) (Jang 1993; Jang and Sur 1995), a fixed structure with grid partition is used. Antecedent and consequent parameters are identified by a combination of LS estimates and gradient-based methods, the so-called *hybrid learning rule*. This method is fast and easy to implement for low-dimensional input spaces. It is more prone to losing transparency and local model accuracy because of the use of *error back-propagation*, which is a global and not locally nonlinear optimization procedure. One possible method to overcome this problem can be to find the antecedents and rules separately, for example, by clustering and constraining the antecedents, and then applying optimization.

Hierarchical NF networks can be used to overcome the dimensionality problem by decomposing the system into a series of MISO and/or SISO systems called *hierarchical systems* (Tachibana and Furuhashi 1994). The local rules use subsets of input spaces and are activated by higher level rules.

The criteria on which NF models are built are based on the requirements for fault diagnosis and the system characteristics. The function of the NF model in the FDI scheme is also important, that is, pre-processing data, identification, (residual generation) or classification (decision making/fault isolation). For example, an NF model with high approximation capability and disturbance rejection is needed for identification, so that the residuals are more accurate, whereas in the classification stage, an NF network with more transparency is required.

1.8.5. *NF model structure identification*

For complexity reduction and transparency, structure identification methods can be applied to find appropriate input partitions, rules and membership functions (MFs). Methods such as Evolutionary Algorithms (EA), Classification and Regression Trees (CART) (Jang 1994), clustering and unsupervised NN (e.g. like the Kohonen feature maps) can be used. Once the structure is determined, that is, the rules and input membership functions, the consequent parameters can be identified by optimization techniques like least-squares estimation. The product space clustering approach can be used (Babuška 1998) for structure identification of TSK and Mamdani fuzzy models. For a MISO nonlinear dynamic system with p inputs, the product space $X \times Y \subset \Re^{p+1}$ is divided into subspaces in which linear models can approximate the nonlinear system. The locally linear model tree LOLIMOT algorithm developed by Nelles and Isermann (1996) can be used to identify a TSK fuzzy model with dynamic linear models as consequent. When using such structure identification techniques, a major issue is the sensitivity to uneven distribution of data. For example, in most clustering algorithms, more clusters are created in regions with more data. A possible solution to this problem may be to initialize the algorithm with a large number of clusters.

The transparency of the NF models can be enhanced by tuning rules and MFs (Babuška 1998). This type of method is referred to as a structure simplification/optimization technique. To find the optimal number of rules, different cluster validity measures and methods like Compatible Cluster Merging (CCM) (Krishnapuram and Freg 1992) can be used. At the NF model level, the rules are further simplified by merging similar fuzzy sets and removing fuzzy sets similar to the universal set. Setnes and Kaymak (1998) used a supervised fuzzy clustering algorithm that uses input–output data, orthogonal techniques and tuning for complexity reduction.

1.8.6. *NF residual generation for FDI*

Figure 1.18 describes an FDI scheme in which several NF models are constructed to identify the faulty and the fault-free behavior of the system.

$$r_i(t) = f\big(u(t), \ldots, u(t-n), y(t), \ldots, y(t-n)\big), \; i = 1, \ldots, m \qquad [1.72]$$

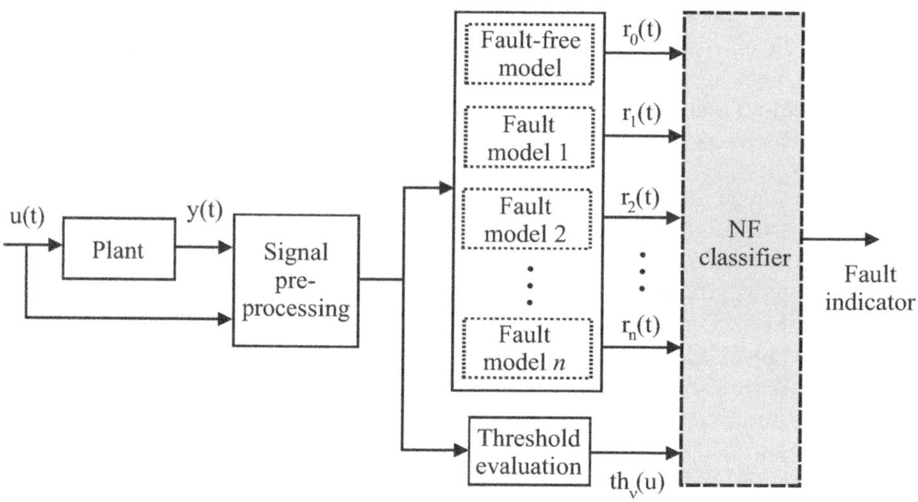

Figure 1.18. *Neuro-fuzzy-based FDI scheme*

Each residual $r_i(t)$ in [1.72] is ideally sensitive to one particular fault in the system. In practice, however, as a consequence of noise and disturbances, residuals are sensitive to more than one fault.

To take into account the sensitivity of residuals to various faults and noise, we apply an NF classifier. A linguistic style (Mamdani) NF network is used that processes the residuals to indicate the fault.

This NF model is constructed with the following set of rules:

$$\textbf{If } r_1 \text{ is small } \ldots r_j \text{ is large, } r_m \text{ is small } \textbf{then } \text{fault}_r \text{ is large} \quad [1.73]$$

Fuzzy threshold evaluation in equation [1.74] is employed to take into account the imprecision of the residual generator at different regions in the input space

$$\text{th}_\nu(u) = \frac{\sum_{i=1}^{C} \text{th}_i \, \eta_i(u)}{\sum_{i=1}^{C} \eta_i(u)} \quad [1.74]$$

where C is the total number of I/P regions with different sensitivity to faults and a multidimensional fuzzy set η_i defines the fuzzy boundary of ith such region. This approach depends heavily on the availability of the faulty and fault-free data, and it is more difficult to isolate faults that appear in the dynamics.

Residuals can also be generated by a nonlinear dynamic model of the plant that approximates a nonlinear dynamic system by local linear models. Such a model can be obtained by *product space clustering* (Babuška 1998) or tree-like algorithms (LOLIMOT algorithm by Nelles and Isermann (1996)). Each local model is a linear approximation of the process in an I/P subspace and the selection of the local model is fuzzy. The output of such a model can be described by:

$$y = \frac{\sum_{i=1}^{C} \alpha_i(u_s) \, f_i}{\sum_{i=1}^{C} \alpha_i(u_s)} \quad [1.75]$$

where f_i is the ith local linear model given by:

$$f_i = \sum_{k=0}^{n} b_{i,k} \, u(t-k) + \sum_{k=0}^{n} a_{i,k} \, y(t-k) + c_i \quad [1.76]$$

$a_{i,k}$, $b_{i,k}$ and c_i are the parameters of the ith model, u_s is the I/P subspace defining the operating point and α_i is the degree to which the ith local model is valid at this operating point.

From $a_{i,k}$, $b_{i,k}$ and c_i, physical parameters such as time constants, static gains, offsets, and so on (Füssel *et al.* 1997) can be extracted for each operating point and can be compared with the parameters estimated online. This approach heavily depends on the accuracy of the nonlinear dynamic model described above. Also, the OE should be minimum when operated in parallel to the system. Moreover, this method requires that there is sufficient excitation at each operating point for the online estimation of parameters. The TSK NF-based FDI scheme is depicted in Figure 1.19.

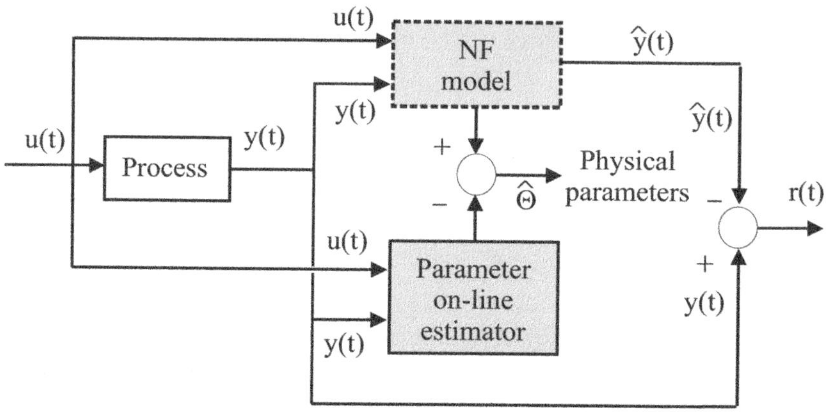

Figure 1.19. *TSK NF-based FDI scheme*

1.9. Conclusion

This chapter has presented a tutorial treatment on the basic principles of model-based FDI.

The FDI problem has been formalized in a uniform framework by presenting a mathematical description and definition. Within this framework, the residual generation has been identified as a central issue in model-based FDI. By choosing the proper design approach, the FDI task can be performed.

The residual generator has been summarized in different residual generation structures. The ways of designing residuals for isolation have also been discussed. The most commonly used residual generation techniques have been introduced by presenting related problems and discussing the applicability of model-based FDI methods.

It is worth noting that the success of fault diagnosis depends on the quality of the residuals. Successful diagnosis requires residual signals that should be robust with respect to modeling uncertainty. The robust FDI problem has also been discussed in this chapter and the implementation of a robust residual generator will be shown in the following chapters.

Other FDI methods such as fuzzy logic, qualitative modeling and NN have been briefly discussed, and the concept of integrated knowledge-based fault diagnosis, utilizing both analytical and heuristic information, has been presented.

1.10. References

Appleby, B., Dowdle, J., Vander Velde, W. (1991). Robust estimator design using μ synthesis. *Proc. of the 30th Conf. on Decision & Control*, Brighton, UK, 640–644.

Ayoubi, M. (1995). Neuro–fuzzy structure for rule generation and application in the fault diagnosis of technical processes. *Proc. of the American Control Conference, ACC'95*, Washington, 2757–2761.

Babuška, R. (1998). *Fuzzy Modelling for Control*. Kluwer Academic Publishers, Dordrecht, The Netherlands.

Basseville, M. (1988). Detecting changes in signals and systems: A survey. *Automatica*, 24(3), 309–326.

Basseville, M. (1997). Information criteria for residual generation and fault detection and isolation. *Automatica*, (33), 783–803.

Basseville, M. and Benveniste, A. (1986). Detection of abrupt changes in signals and dynamical systems. *Lecture Notes in Control and Information Sciences*, Allgöwer, F. and Morari, M. (eds). Springer–Verlag, London.

Basseville, M. and Nikiforov, I.V. (1993). *Detection of Abrupt Changes: Theory and Application*. Prentice Hall, Hoboken, NJ.

Beard, R.V. (1971). Failure accomodation in linear systems through self-reorganisation. Technical Report MVT-71-1, Man Vehicle Lab., Cambridge, MA.

Boyd, S., Ghaoui, L., Feron, E., Balakrishnan, V. (1994). *Linear Matrix Inequalities in System and Control Theory*. SIAM, Philadelphia.

Brown, M. and Harris, C. (1994), *Neurofuzzy Adaptive Modelling and Control*. Prentice Hall, Hemel Hempstead, UK.

Calado, J., Korbicz, J., Patan, K., Patton, R.J., Sá da Costa, J. (2001). Soft computing approaches to fault diagnosis for dynamic systems. *European Journal of Control*, 7(2–3), 248–286.

Carpenter, G. and Grossberg, S. (1987). A massively parallel architecture for a self-organizing neural pattern recognition machine. *Computer Vision, Graphics and Image Processing*, 37, 54–115.

Chen, J. and Patton, R.J. (1999). *Robust Model-Based Fault Diagnosis for Dynamic Systems*. Kluwer Academic, Publishers, Norwell, MA.

Chen, J. and Patton, R.J. (2000). Standard H_∞ filter formulation of robust fault detection. In *SAFEPROCESS2000, 4th IFAC Symposium on Fault Detection, Supervision and Safety for Technical Processes*, Edelmayer, AM. (ed.). Elsevier Ltd., London.

Chen, J. and Patton, R.J. (2001). Fault-tolerant control systems design using the linear matrix inequality method. *European Control Conference, ECC'01*, Porto, Portugal, 1993–1998.

Chen, R.H. and Speyer, J.L. (1999). Residual-sensitive fault detection filter. In *Proceedings of the 7th IEEE Mediterranean Conference on Control and Automation, MED'99*, Palmor, Z.J. (ed.). Haifa, Israel, Elsevier Ltd., London.

Chen, S., Billings, A.S., Cowan, C.F.N., Grant, P.M. (1990). Practical identification of NARMAX models using radial basis function. *Int. J. Control*, 52, 1327–1350.

Chen, J., Patton, R.J., Zhang, H.Y. (1993). A multi-criteria optimization approach to the design of robust fault detection algorithm. *Proccedings of Int. Conf. on Fault Diagnosis: TOOLDIAG'93*, Toulouse, France.

Chen, J., Patton, R.J., Liu, G.P. (1996a). Optimal residual design for fault-diagnosis using multiobjective optimisation and genetic algorithms. *Int. J. Sys. Sci.*, 27(6), 567–576.

Chen, J., Patton, R.J., Zhang, H.Y. (1996b). Design of unknown input observer and robust fault detection filters. *Int. J. Control*, 63(1), 85–105.

Chen, Z., Patton, R.J., Chen, J. (1997). Robust fault-tolerant system synthesis via LMI. In *Proccedings of IFAC Symposium on Fault Detection, Supervision and Safety for Technical Processes: SAFEPROCESS'97*, Staroswiecki, M. (ed.). University of Hull, UK.

Chiang, L.H., Russel, E.L., Braatz, R.D. (2001). *Fault Detection Diagnosis in Industrial Systems*. Springer-Verlag London Limited, London.

Chow, E.Y. and Willsky, A.S. (1980). Issue in the development of a general algorithm for reliable failure detection. *Proceedings of the 19th Conf. on Decision & Control*, Albuquerque, NM.

Chow, E.Y. and Willsky, A.S. (1984). Analytical redundancy and the design of robust detection systems. *IEEE Trans. Automatic Control*, 29(7), 603–614.

Chung, W.H. and Speyer, J.L. (1998). A game theoretic fault detection filter. *IEEE Trean. on Automatic Control*, 43(2), 143–161.

Clark, R.N. (1978). Instrument fault detection. *IEEE Trans. Aero. & Electronic Systems*, 14(3), 455–465.

Clark, R.N. (1989). Model-based methods for fault diagnosis: Some guide-lines. *Fault Diagnosis in Dynamic Systems: Theory and Application*. Prentice Hall, Upper Saddle River, NJ.

Daly, K.C., Gai, E., Harrison, J.V. (1979). Generalized likelihood test for FDI in redundancy sensor configurations. *J. of Guidance, Control & Dynamics*, 2(1), 9–17.

Davis, L. (1991). *Handbook of Genetic Algorithms*. Van Nostrand Reinhold, New York.

Delmaire, G., Cassar, P., Staroswiecki, M., Christophe, C. (1999). Comparison of multivariable identification and parity space techniques for FDI purpose in M.I.M.O. systems. *European Control Conference, ECC'99*, Karlsruhe, Germany.

Dexter, A.L. and Benouarets, M. (1997). Model-based fault diagnosis using fuzzy matching. *IEEE Trans. on Sys. Man. and Cyber. Part A: Sys. & Humans*, 27(5), 673–682.

Dietz, W.E., Kiech, E.L., Ali, M. (1989). Jet and rocket engine fault diagnosis in real time. *J. of Neural Network Computing*, 1, 5–18.

Ding, X. and Frank, P.M. (1990). Fault detection via factorization approach. *Syst. Contr. Lett.*, 14(5), 431–436.

Ding, X. and Frank, P.M. (1991). Frequency domain approach and threshold selector for robust model-based fault detection and isolation. *IFAC/IMACS Symposium SAFEPROCESS'91*, vol. 1, Baden-Baden, 307–312.

Ding, S.X., Jeinsch, T., Ding, E.L., Zhou, D., Wang, G. (1999). Application of observer-based FDI schemes to the three tank system. *European Control Conference, ECC'99*, Karlsruhe, Germany.

Ding, S.X., Jeinsch, T., Frank, P.M., Dind, E.L. (2000). A unified approach to the optimisation of fault detection systems. *Int. J. of Adaptive Control and Signal Processing*, 14(7), 725–745.

Duan, G., How, D., Patton, R.J. (2002). Robust fault detection in descriptor systems via generalised unknown input observers. *Int. J. Systems Science*, 32(2), 7724–7729.

Edelmayer, A., Bokor, J., Keviczky, L. (1997). A scaled L_2 optimisation approach for improving sensitivity of $H_inf ty$ detection filters for LTV systems. In *2nd IFAC Symp. on Robust Control Design: RECOND97*, Bányász, C. (ed.). Elsevier Ltd., London.

Emami-Naeini, A., Akhter, M., Rock, M. (1988). Effect of model uncertainty on failure detection: The threshold selector. *IEEE Trans. on Automatic Control*, 33(2), 1106–1115.

Frank, P.M. (1993). Advances in observer-based fault diagnosis. *Proceedings TOOLDIAG'93 Conference*, CERT, Toulose, France.

Frank, P.M. (1994a). Application of fuzzy logic process supervision and fault diagnosis. In *SAFEPROCESS'94: Preprints of the IFAC Symposium on Fault Detection, Supervision and Safety for Technical Processes*, vol. 2, Espoo, Finland, Elsevier Ltd., London.

Frank, P.M. (1994b). Enhancement of robustness on observer-based fault detection. *International Journal of Control*, 59(4), 955–983.

Frank, P.M. (1996). Analytical and qualitative model-based fault diagnosis: A survey and some new results. *European Journal of Control*, 2(1), 6–28.

Frank, P.M. and Ding, X. (1997). Survey of robust residual generation and evaluation methods in observer-based fault detection system. *Journal of Process Control*, 7(6), 403–424.

Frank, P.M., Ding, S.X., Köpper-Seliger, B. (2000). Current developments in the theory of FDI. In *SAFEPROCESS'00: Preprints of the IFAC Symposium on Fault Detection, Supervision and Safety for Technical Processes*, vol. 1, Budapest, Hungary, Elsevier Ltd., London.

Füssel, D., Ballé, P., Isermann, R. (1997). Closed loop fault diagnosis based on a non-linear process model and automatic fuzzy rule generation. *Proccedings of IFAC Symposium on Fault Detection, Supervision and Safety for Technical Process SAFEPROCESS'97*, University of Hull, UK.

Gertler, J. (1991). Generating directional residuals with dynamic parity equations. *Proceedings of the IFAC/IMACS Symp. SAFEPROCESS'91*, Baden-Baden, Germany.

Gertler, J. (1995). Diagnosing parametric faults: From parameter estimation to parity relations. *American Control Conference, ACC'95*, IEEE Piscataway, Seattle, Washington, NJ, 1615–1620.

Gertler, J. (1998). *Fault Detection and Diagnosis in Engineering Systems*. Marcel Dekker, New York.

Gertler, J. and Monajemy, R. (1993). Generating directional residuals with dynamic parity equations. *Proceedings of the 12th IFAC World Congress*, 7, 505–510.

Gertler, J. and Singer, D. (1990). A new structural framework for parity equation-based failure detection and isolation. *Automatica*, 26(2), 381–388.

Goldberg, D.E. (1989). *Genetic Algorithms in Search, Optimisation and Machine Learning*. Addison Wesley Publishing Company, Boston, MA.

Guidorzi, R.P. (1975). Canonical structures in the identification. *Automatica*, 11, 361–374.

Himmelblau, D.M. (1978). *Fault Diagnosis in Chemical and Petrochemical Processes*. Elsevier, Amsterdam.

Himmelblau, D.M., Barker, R.W., Suewatanakul, W. (1991). Fault classification with the aid of artificial neural networks. In *IFAC/IMACS Symposium SAFEPROCESS '91*, vol. 2, Baden-Baden, Germany, Elsevier Ltd., London.

Hoskins, J.C. and Himmelblau, D.M. (1988). Artificial neural network models of knowledge representation in chemical engineering. *Comp. Chem. Engng*, 12, 881–890.

Hou, M. and Patton, R.J. (1996). An LMI approach to H_-/H_∞ fault detection observers. In *UKACC International Conference on Control, CONTROL'96*, University of Exeter, UK.

Hou, M. and Patton, R.J. (1997). An H_∞/H_- approach to the design of robust fault diagnosis observers based upon LMI optimisation. *Proceedings of the 4th European Control Conference, ECC'97*, Brussels.

Isermann, R. (1984). Process fault detection based on modelling and estimation methods: A survey. *Automatica*, 20(4), 387–404.

Isermann, R. (1992). Estimation of physical parameters for dynamic processes with application to an industrial robot. *Int. J. of Control*, 55, 1287–1298.

Isermann, R. (1993). Fault diagnosis via parameter estimation and knowledge processing. *Automatica*, 29(4), 815–835.

Isermann, R. (1994). Integration of fault detection and diagnosis methods. *Proceedings of the IFAC SAFEPROCESS Symposium '94*, Espoo, Finland.

Isermann, R. (1997). Supervision, fault detection and fault diagnosis methods: An introduction. *Control Engineering Practice*, 5(5), 639–652.

Isermann, R. (1998). On fuzzy logic applications for automatic control, supervision and fault diagnosis. *IEEE Trans. on Sys. Man. and Cyber. Part A: Sys. & Humans*, 28(2), 221–235.

Isermann, R. and Ballé, P. (1997). Trends in the application of model-based fault detection and diagnosis of technical processes. *Control Engineering Practice*, 5(5), 709–719.

Isermann, R. and Freyermuth, B. (eds). (1992). *Fault Detection, Supervision and Safety for Technical Processes, SAFEPROCESS'91*. Pergamon Press, Oxford, UK.

Jang, J. (1993). ANFIS: Adaptive network based fuzzy inference system. *IEEE Transactions on Systems, Man., & Cybernetics*, 23(3), 665–684.

Jang, J. (1994). Structure determination in fuzzy modelling: A fuzzy CART approach. *Proceedings of IEEE International Conference on Fuzzy Systems*, Orlando, FL, USA.

Jang, J. and Sur, R. (1995). Neuro–fuzzy modelling and control, *Proc. IEEE*, 83(3), 378– 405.

Jazwinski, A.H. (1970). *Stochastic Processes and Filtering Theory*. Academic Press, New York.

Jones, H.L. (1973). Failure detection in linear systems. PhD Thesis. Department of Aeronautics, M.I.T., Cambridge, MA.

Kalman, R.E. (1982). System identification from noisy data. In *Dynamical System II*, Bednarek, A.R., Cesari, L. (eds). Academic Press, New York.

Kalman, R.E. (1990). Nine lectures on identification. *Lecture Notes on Economics and Mathematical System*.

Korbicz, J., Patan, K., Obuchowicz, A. (1999). Dynamic neural network for process modelling in fault detection and isolation systems. *Applied Mathematics and Computer Science*, Technical University of Zielona Gora, Poland, 9(2), 519–546.

Kramer, M.A. (1987). Malfunction diagnosis using quantitative models with non-boolean reasoning in expert systems. *AIChE*, (33), 130–140.

Krishnapuram, R. and Freg, C. (1992). Fitting an unknown number of lines and planes to image data through compatible cluster merging. *Pattern Recognition*, 25(4), 385–400.

Kuo, B.C. (1995). *Automatic Control Systems*, 7th edition. Prentice Hall, Englewood Cliffs, NJ.

Leontaritis, I. and Billings, S.A. (1985a). Input-output parametric models for nonlinear systems part II: Stochastic non-linear systems. *Int. J. Control*, 41(2), 329–344.

Leontaritis, I. and Billings, S.A. (1985b). Input-output parametric models for non-linear systems part I: Deterministic non-linear systems. *Int. J. Control*, 41(2), 303–328.

Liu, G.P. and Patton, R.J. (1998). *Eigenstructure Assignment for Control System Design*. John Wiley & Sons, Chichester, UK.

Ljung, L. (1999). *System Identification: Theory for the User*, 2nd edition. Prentice Hall, Englewood Cliffs, NJ.

Lou, X., Willsky, A., Verghese, G. (1986). Optimal robust redudancy relations for failure detection in uncertainty systems. *Automatica*, 22(3), 333–344.

Mamdani, E. (1976). Advances in the linguistic synthesis of fuzzy controllers. *Int. J. Man–Machine Studies*, 8, 669–678.

Mamdani, E. and Assilian, S. (1995). An experiment in linguistic synthesis with fuzzy logic controller. *Int. J. Man–Machine Studies*, 7(1), 1–13.

Mangoubi, R., Appleby, B.D., Farrell, J.R. (1992). Robust estimation in fault detection. In *Proceedings of the 31st Conference on Decision & Control*, IEEE, Piscataway, NJ, 2317–2322.

Massoumnia, M.A. (1986). A geometric appoach to failure detection and identification in linear systems. PhD Thesis, M.I.T., Cambridge, MA.

Massoumnia, M.A., Verghese, G.C., Willsky, A.S. (1989). Failure detection and identification. *IEEE Trans. Automat. Contr.*, 34, 316–321.

McDuff, R.J. and Simpson, P.K. (1990). An adaptive reasonance diagnostic system. *Journal of Neural Network Computing*, (2), 19–29.

Meneganti, M., Saviello, F., Tagliaferri, R. (1998). Fuzzy neural networks for classification and detection of anomalies. *IEEE Trans. on Neural Networks*, 9(5), 848–861.

Nelles, O. (2001). *Nonlinear System Identification*. Springer-Verlag, Berlin, Germany.

Nelles, O. and Isermann, R. (1996). Basis function networks for interpolation of local linear models. In *Proceedings of the 35th IEEE Conference on Decision and Control*, vol. 4, Kobe, Japan.

Niemann, H. and Stoustrup, J. (1996). Filter design for failure detection and isolation in the presence of modelling errors and disturbances. In *Proceedings of the 35th IEEE Conference on Decision and Control*, Kobe, Japan, 1155–1160.

Norton, J. (1986). *An Introduction to Identification*. Academic Press, London.

Patton, R.J. and Chen, J. (1991a). A review of parity space approaches to fault diagnosis. *IFAC Symposium SAFEPROCESS '91*, Baden-Baden, Germany.

Patton, R.J. and Chen, J. (1991b). Robust fault detection using eigenstructure assignment: A tutorial consideration and some new results. *30th IEEE Conference on Decision and Control*, IEEE Piscataway, NJ, 2242–2247.

Patton, R.J. and Chen, J. (1993). Optimal selection of unknown input distribution matrix in the design of robust observers for fault diagnosis. *Automatica*, 29(4), 837–841.

Patton, R.J. and Chen, J. (1994). A review of parity space approaches to fault diagnosis for aerospace systems. *AIAA J. of Guidance, Contr. & Dynamics*, 17(2), 278–285.

Patton, R.J. and Chen, J. (1997). Observer-based fault detection and isolation: Robustness and applications. *Control Eng. Practice*, 5(5), 671–682.

Patton, R.J. and Chen, J. (2000). On eigenstructure assignment for robust fault diagnosis. *Int. J. of Robust & Non-Linear Control*, 10(9), 1193–1208.

Patton, R.J. and Hou, M. (1997). h_∞ estimation and robust fault detection. *Proceedings of the 1997 European Control Conference, ECC'97*, Brussels, Belgium.

Patton, R.J., Willcox, S., Winter, J. (1986). A parameter insensitive technique for aircraft sensor fault diagnosis using eigenstructure assignment and analytical redundancy. *Proc. of the AIAA Conference on Guidance, Navigation & Control*, number 86-2029-CP, Williamsburg, VA.

Patton, R.J., Frank, P.M., Clark, R.N. (eds) (1989). *Fault Diagnosis in Dynamic Systems, Theory and Application*. Prentice Hall, London.

Patton, R.J., Lopez-Toribio, C.J., Uppal, F.I. (1999). Artificial intelligence approaches to fault diagnosis. *Applied Mathematics and Computer Science*, 9(3), 471–518.

Patton, R.J., Frank, P.M., Clark, R.N. (eds) (2000). *Issues of Fault Diagnosis for Dynamic Systems*. Springer-Verlag, London.

Pau, L.F. (1981). *Failure Diagnosis and Performance Control*. Marcel Dekker, New York.

Potter, I.E. and Suman, M.C. (1977). Thresholdless redundancy management with array of skewed instruments. Technical Report AGARDOGRAPH-224, AGARD, Integrity in Electronic Flight Control Systems.

Ray, A. and Luck, R. (1991). An introduction to sensor signal validation in redundant measurement systems. *IEEE Contr. Syst. Mag.*, 11(2), 44–49.

Rich, S.H. and Venkatasubramanian, V. (1987). Model-based reasoning in diagnostic expert system for chemical process plant. *Comp. Chem. Engng*, 11, 111–122.

Sadrnia, M.A., Chen, J., Patton, R.J. (1997). Robust H_∞/μ observer-based residual generation for fault diagnosis. *Proceedings of the IFAC Symp. on Fault Detection, Supervision and Safety for Technical Processes: SAFEPROCESS'97*, University of Hull, UK and Pergamon Press, Oxford, UK, 155–162.

Sauter, D., Rambeaux, F., Hamelin, F. (1997). Robust fault diagnosis in a H_∞ setting. *Proceedings of the IFAC Symp. on Fault Detection, Supervision and Safety for Technical Processes: SAFEPROCESS'97*, University of Hull, UK and Pergamon Press, Oxford, UK, 867–874.

Setnes, M. and Kaymak, U. (1998). Extended fuzzy c–means with volume prototypes and cluster merging. In *Proceedings EUFIT98*, vol. 3, Aachen, Germany, 1360–1364.

Söderström, T. and Stoica, P. (1987). *System Identification*. Prentice Hall, Englewood Cliffs, NJ.

Stoustrup, I. and Niemann, H. (1998). Fault detection for nonlinear systems – A standard problem approach. *Proceedings of the 37th IEEE Conf. on Decision & Control*, IEEE, Piscataway, NJ, 96–101.

Stoustrup, J., Grimble, M.J., Niemann, H. (1997). Design of integrated systems for the control and detection of actuator/sensor faults. *Sensor Review*, 17(2), 138–149.

Sugeno, M. and Kang, G. (1988). Structure identification of fuzzy model. *Fuzzy Set and Systems*, 28, 15–33.

Tachibana, K. and Furuhashi, T. (1994). A hierarchical fuzzy modelling method using genetic algorithm for identification of concise submodels. *Proc. of 2nd Int. Conference on Knowledge – Based Intelligent Electronic Systems*, Adelaide, Australia.

Takagi, T. and Sugeno, M. (1985). Fuzzy identification of systems and its application to modelling and control. *IEEE Transaction on System, Man and Cybernetics*, SMC-15(1), 116132.

Tou, J.T. and Gonzalez, R.C. (1974). *Pattern Recognition Principles*. Addison Wesley Publishing Company, Boston, MA.

Ulieru, M. and Isermann, R. (1993). Design of fuzzy-logic based diagnostic model for technical process. *Fuzzy Set and Systems*, 58(3), 249–271.

Uppal, F.J. and Patton, R.J. (2000). Application of B-spline neuro-fuzzy networks to identification, fault detection and isolation. In *SAFEPROCESS 2000, 4th Symp. on Fault Detection, Supervision and Safety for Technical Process*, vol. 1, Budapest, Hungary, Elsevier Ltd., London.

Venkatasubramanian, V. and Chan, K. (1989). A neural network methodology for process fault diagnosis. *AIChE J.*, 35, 1993–2002.

Watanabe, K. and Himmelblau, D.M. (1982). Instrument fault detection in systems with uncertainties. *Int. J. System Sci.*, 13(2), 137–158.

Willsky, A.S. (1976). A survey of design methods for failure detection in dynamic systems. *Automatica*, 12(6), 601–611.

Wu, Z.Q. and Harris, C.J. (1996). Neuro-fuzzy modelling and state estimation. In *IEEE Medit. Symp. on Control and Automation: Circuits, Systems and Computers '96*. Hellenic Naval Academy, Piraeus, Greece, 603–610.

Wünnenberg, J. (1990). Observer-based fault detection in dynamic systems. PhD Thesis, University of Duisburg, Duisburg, Germany.

Wünnenberg, J. and Frank, P.M. (1987). Sensor fault detection via robust observer. In *System Fault Diagnosis, Reliability, and Related Knowledge-Based Approaches*, Tzafestas, S., Singh, M., Schmidt, G. (eds). Springer, Cham, Switzerland.

Wünnenberg, J. and Frank, P.M. (1990). Robust observer-based detection for linear and nonlinear systems with application to robot. *Proc. of IMACS Annals on Computing & Applied Mathematics MIM–*S^2*:90*, Brussels.

Zhou, K., Doyle, J.C., Glover, K. (1996). *Robust and Optimal Control*. Prentice Hall, Upper Saddle River, NJ.

2

Structural Analysis

Mattias KRYSANDER and Erik FRISK
Department of Electrical Engineering, Linköping University, Sweden

2.1. Introduction

As illustrated in Chapter 1, there are many methods available for designing diagnosis systems based on the mathematical description of the system under supervision. This chapter will further that discussion to include fault diagnosis analysis of models since, in addition to designing diagnosis systems – that is, detection algorithms and fault isolability approaches – it is also of interest to analyze models and determine relevant fault diagnosis properties. For example, determining which faults can be detected, which faults can be isolated and where to put sensors.

The complexity of the analysis and design increases with model complexity. For linear time-invariant models, rather complete and exhaustive results can be obtained (Nyberg and Frisk 2006). However, as model complexity increases, for example, including nonlinear constraints in the model description, more advanced analytical techniques are needed, which may not scale well to large-scale models.

A useful approach to analysis and design for large-scale, nonlinear problems is structural analysis, the topic of this chapter. Structural analysis utilizes only coarse model information which means large problems can be analyzed efficiently and without numerical difficulties; the cost, however, is that results are generic or best-case. Structural analysis can be used to assess detectability and isolability properties of a model, determine where to put sensors to achieve required performance and also to aid the design of nonlinear residual generators and fault detectors. Structural analysis for diagnosis has been proposed in numerous works

Diagnosis and Fault-tolerant Control 1,
coordinated by Vicenç PUIG and Silvio SIMANI.

(Krysander *et al.* 2008; Cassar and Staroswiecki 1997; Ploix *et al.* 2005; Pulido and González 2004). Structural analysis is well suited to computer support, and all examples in this chapter have been performed using the freely available toolbox (Frisk *et al.* 2017), available both in MATLAB and Python. The toolbox is easily accessible[1] and the MATLAB version can simply be downloaded. The Python version can be pip-installed from the standard Python package index. Code, MATLAB and Python, corresponding to the examples in this chapter, can be found online[2].

In this chapter, a formal definition of fault detection and isolation is given in terms of consistency relations. Then, the definitions are made generic, which are suitable for structural analysis using graph theoretical algorithms. The chapter starts by introducing structural models and basic algorithms on graphs. Then, techniques for diagnosability analysis of models and efficient algorithms for finding all testable submodels in a system are explained.

2.2. Background

This section will introduce basic concepts for structural analysis. First, we define how the structure of a model can be represented by a graph. Then a graph-theoretic algorithm called Dulmage–Mendelsohn decomposition is introduced, which is fundamental for analyzing diagnostic properties of models.

2.2.1. *Structural models*

A system consists of a set of components and sensors. Its dynamic behavior is described by a model, that is, a set of constraints that describes the behavior and relations among all model variables. The formal expression of the model depends on the type of knowledge (analytical, qualitative, production rules or numerical tables, etc.). It generally relies on a component-based description, which associates a set of constraints to each component. The structural analysis described later in this chapter will use a structural representation of the model. As mentioned previously, the structural representation is a coarse model description that only considers which variables are included in each model relation/equation.

Let E be the set of constraints or equations in the model and V the set of variables, then the structural model can be represented by the bipartite graph $G = (E \bigcup V,\ A)$, where A is a set of edges between nodes in the two node sets E and V. An edge $(e_i, v_j) \in A$ if and only if variable $v_j \in V$ appears in the model relation $e_i \in E$. A

1 https:// faultdiagnosistoolbox.github.io/.

2 https://gitlab.liu.se/erifr93/structuralanalysischapter/.

common way to visualize a structural model is with the so-called *bi-adjacency matrix* of the graph. The bi-adjacency matrix is a matrix with equally as many rows and columns as constraints and variables, respectively. An element in position (i, j) of the bi-adjacency matrix is empty when the variable v_j is not included in constraint e_i. The bi-adjacency matrix of a graph representing the structure of a model will also simply be called the structural matrix.

The set of variables can be classified as unknown, $\mathcal{X} \subset V$, known, $\mathcal{Z} \subset V$, or fault, $\mathcal{F} \subset V$ being $V = \mathcal{X} \bigcup \mathcal{Z} \bigcup \mathcal{F}$. For analysis purposes, the part of the structure related to unknown variables will be most important, and the subgraph including only constraints and unknown variables will be called the *reduced structure graph*.

A small electric motor system will be used as a running example to exemplify concepts and algorithms, although the results will be equally applicable to large-scale, nonlinear differential-algebraic models. Figure 2.1 shows an equivalent circuit representation of the electric motor. It is a series-wound direct current (DC) motor with terminal voltage V as the control input delivering a motor torque T_m. The output shaft is connected to an unknown load torque T_l. The current i, the shaft torque T and the angular velocity ω are measured with the corresponding measurement signals y_i, y_T and y_ω, respectively. The motor is designed with a nominal winding resistance R, inductance L and the electromechanical coupling is expressed by the constant K_f. The faults considered are sensor faults of the all sensors, that is, f_i, f_ω and f_T, a resistance fault f_R and a fault acting as an additional load f_{load}. The faults are assumed to be zero in the fault-free case and non-zero otherwise. This means that the fault models are general in the sense that they are valid for faults of different sizes and dynamic behaviors.

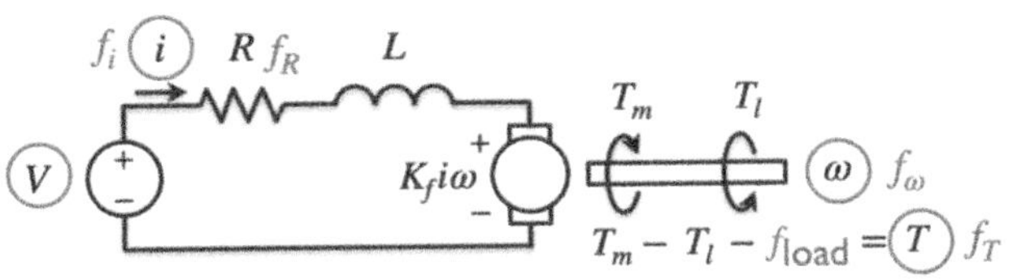

Figure 2.1. *A circuit view of the electric motor that will be used as a running example. For a color version of this figure, see www.iste.co.uk/puig/diagnosis1.zip*

The equations describing the behavior of the motor are given as

$$e_1 : V = i(R + f_R) + Li' + K_f i\omega \qquad e_6 : y_\omega = \omega + f_\omega$$

$$e_2 : i = \sqrt{\frac{T_m}{K_f \omega}} \qquad e_7 : y_T = T + f_T$$

$$e_3 : J\omega' = T - b\omega \qquad e_8 : \frac{di}{dt} = i'$$

$$e_4 : T = T_m - T_l - f_{\text{load}} \qquad e_9 : \frac{d\omega}{dt} = \omega' \qquad [2.1]$$

$$e_5 : y_i = i + f_i$$

where the combined rotational inertia of the rotor and load is J and the friction coefficient is b. This model is a simple motor model not valid at magnetic saturation. Equation e_1 is derived from Kirchhoff's voltage law applied to the circuit, e_2 is derived from the torque equation, e_3 is derived from Newton's second law of motion relating angular velocity ω of the motor with the net torque T and e_4 is derived from the torque balance. Equations e_5–e_7 are the sensor equations including sensor faults. Equations e_8 and e_9 describe the dynamics, that is, define that the time derivative of x is equal to x'. In summary, the unknown signals are $\mathcal{X} = \{i, i', \omega, \omega', T, T_m, T_l\}$, the known signals, i.e. control inputs and sensor signals, are $\mathcal{Z} = \{y_i, y_\omega, y_T, V\}$, the fault signals are $\mathcal{F} = \{f_R, f_i, f_\omega, f_T, f_{\text{load}}\}$ and the known parameters are $\{R, L, K_a, b, J\}$.

Note that all dynamics of the model are separated from the algebraic part by writing the dynamics with so-called differential constraints e_8 and e_9 in the particular form

$$\frac{dx}{dt} = x' \quad \text{for } x \in \{i, \omega\}$$

This way of representing dynamics is useful in structural analysis. In the structural model, special entries in the matrix indicate a differential constraint, a D for x', that is, if solving for this variable differentiation of x is required, and I for x because solving this variable requires integration of x'. A structural representation of the motor model [2.1] is shown in Figure 2.2. The blue, red and black parts correspond to unknown, fault and known variables, respectively. The dots, for example, on row e_1 mean that e_1 involves the known variables i, i' and ω, the fault f_R and the known variable V. The differential constraints e_8 and e_9 are identified by the special marks I and D in the structural matrix.

2.2.2. *Dulmage–Mendelsohn decomposition and matchings*

A key tool when analyzing structural models is the Dulmage–Mendelsohn decomposition (Dulmage and Mendelsohn 1958) used for diagnosis in Blanke *et al.* (2006) and Krysander *et al.* (2008). The general Dulmage–Mendelsohn decomposition is illustrated in Figure 2.3 where, by a suitable reordering of constraints and variables, the bi-adjacency matrix of a reduced structural graph is converted to a block triangular form.

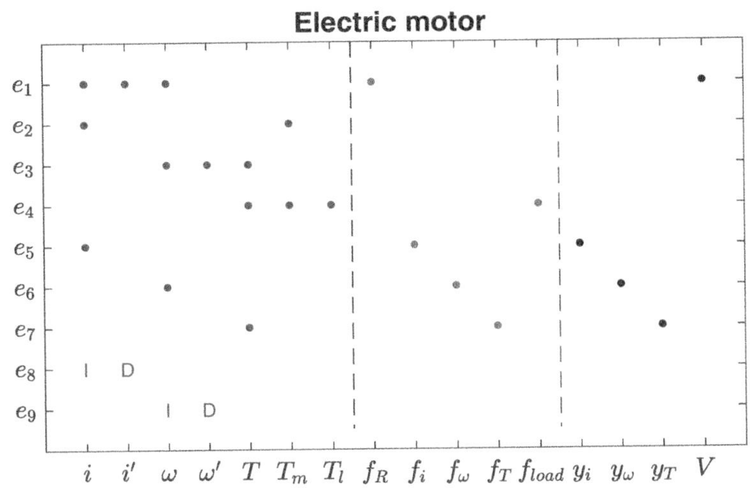

Figure 2.2. *Structural model of the electric motor*

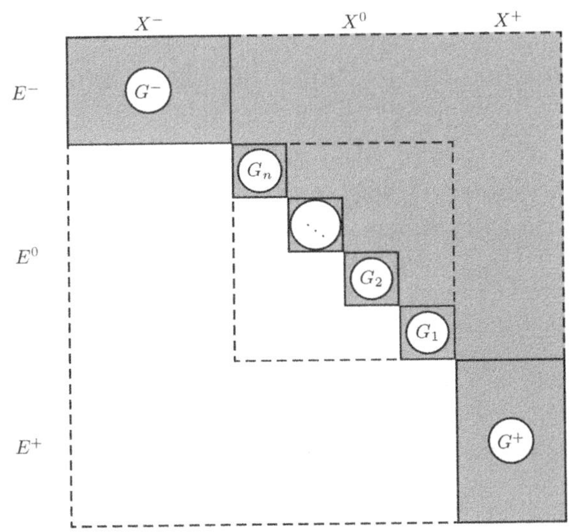

Figure 2.3. *Dulmage–Mendelsohn decomposition with over- and underdetermined part, the just determined part with the finer structure in the Hall components*

The subgraph G^- with node sets E^- and X^- represents the underdetermined part of the model, G^0 with node sets E^0 and X^0 represents the exactly determined part and G^+ with node sets E^+ and X^+ represents the overdetermined part. The overdetermined part contains redundancy and can therefore be used for diagnosis. In the exactly determined part, there is a finer structure of Hall components, which is denoted as G_i. Note that the Dulmage–Mendelsohn components $G^-, G_1, \ldots G_n, G^+$ are unique, but the triangular structure may vary, for example if the Hall components are independent. The notation $()^+$ will be used as an operator with a slight abuse of notation.

A key concept used frequently in the following sections is *matching*.

DEFINITION 2.1 (Matching).– *For a bi-partite graph $G = (E \bigcup V, A)$, a matching $\Gamma \subset A$ is a set of edges with no common vertices. A matching is said to be* maximal *if it contains the largest number of edges. A matching is said to be* perfect *if every vertex in the graph is matched, that is, it is included in an edge in Γ.*

Matchings have many uses in fault diagnosis, which will be explored in the following sections, for example the structural rank of a matrix equals the size of a maximal matching. Fundamental matrix properties and the relation to structural properties are described in Murota (2009). Also, in this context, a matching can be interpreted as if equation e_i is matched with the variable x_j, then e_i can be interpreted as a mechanism for solving x_j if all other variables are known. This means that a matching can be useful for finding a causal interpretation of the model, that is, a computational order for the variables that will be further explored in section 2.4.3. Note that there are multiple complete matchings in an overdetermined system.

Figure 2.4(a) shows the Dulmage–Mendelsohn decomposition of the reduced structure graph of the motor model in Figure 2.2. The decomposition shows that there is no underdetermined part, and the exactly determined part is $\{e_2, e_4\}$ making it possible to solve for $\{T_l, T_m\}$ and the remaining equations in the rectangular block are the overdetermined part that contain all redundancy. The red diagonal indicates a maximal matching, which is a partial result needed for computing the decomposition. The equations involved in the matching are sufficient in number to solve for all variables. The two equations that are not part of the matching, that is, e_5 and e_6 are so-called redundant relations that can be used for diagnosis.

DEFINITION 2.2 (Redundant Relation).– *Let a bipartite graph $G = (E \bigcup \mathcal{X}, A)$ represent the structure of a model and $\Gamma \subseteq A$ a maximal matching in G. Then a Redundant Relation (RR) is a relation in E2+, which is not involved in the matching Γ.*

Redundancy relations are not needed to determine any of the unknown variables. Every RR produces an *Analytical Redundant Relation* (ARR), when the unknown variables involved in the RR are replaced by their expression involving only known variables, as will be seen in section 2.4.

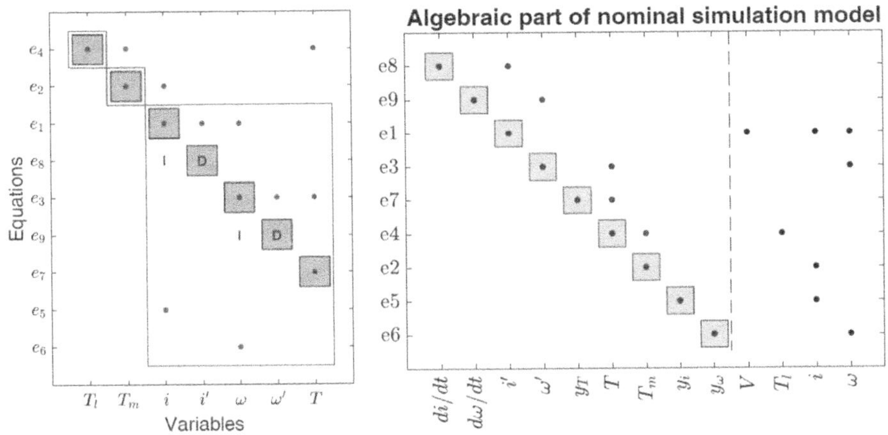

(a) Decomposition of diagnosis model

(b) Decomposition of simulation model

Figure 2.4. *Dulmage–Mendelsohn's decomposition applied to variants of the electric motor model. For a color version of this figure, see www.iste.co.uk/puig/diagnosis1.zip*

2.2.3. *Dulmage–Mendelsohn decomposition and simulation*

Structural methods can also aid simulation of differential-algebraic equations (DAEs) and especially models in a semi-explicit form

$$\begin{aligned} \dot{x}_1 &= f(x_1, x_2, u) \\ 0 &= g(x_1, x_2, u) \end{aligned} \qquad [2.2]$$

where x_1 is a vector of the state variables, x_2 is a vector of the algebraic variables, u are the inputs, and $\dot{x}$ is the time-derivative of a variable x. A DAE solver needs a function to compute $\dot{x}_1$ and possibly also x_2 given values of x_1 and u, that is, to consider $\mathcal{X} = \{\dot{x}_1, x_2\}$ as unknowns and $\mathcal{Z} = \{u, x_1\}$ as known variables. Dulmage–Mendelsohn decomposition of the graph defined by $\mathcal{X} = \{\dot{x}_1, x_2\}$ defines a sequential computation scheme for $\dot{x}_1$ and x_2, as will be illustrated in the next example.

Assume that the fault-free electric motor should be simulated with the voltage V and load torque T_l as inputs. In this case, the electric motor model is in the semi-explicit form [2.2] where faults are zero, $u = (V, T_l)$, $x_1 = (i, \omega)$, and $x_2 = (i', \omega', T, T_m, y_i, y_\omega, y_T)$. A sequential computation scheme for $\dot{x}_1$ and x_2 is obtained by applying Dulmage–Mendelsohn decomposition to the structure matrix, where $\dot{x}_1$ and x_2, are considered to be unknown, as shown in Figure 2.4(b). It is exactly determined and the matching in red defines a sequential way of computing $\dot{x}_1$

and x_2 by starting from the bottom and solve for the matched variable in each equation, respectively. Then we get the following computation scheme:

$$\begin{aligned} &e_6 : y_\omega = \omega && e_3 : \omega' = (T - b\omega)/J \\ &e_5 : y_i = i && e_1 : i' = (V - i(R + f_R) - K_f i\omega)/L \\ &e_2 : T_m = K_f \omega i^2 && e_9 : \dot{\omega} = \omega' \\ &e_4 : T = T_m - T_l && e_8 : \dot{i} = i' \\ &e_7 : y_T = T && \end{aligned}$$

In this example, the matrix is purely upper triangular, and the solution is obtained by solving for one variable at a time. In general, there might be diagonal blocks, see, for example, G_i in Figure 2.3, of size greater than one and the equations and variables included in such blocks must be solved simultaneously as a system of equations. Sequential computation utilizing structural properties will be further developed for residual generation in section 2.4.

2.3. Fault isolability analysis

The set of faults that are detectable and isolable from one another in a system depend on the physics of the system and its interconnections, as well as the measurements available. Fault detectability and isolability are system properties that put limits on the diagnosis performance, which can be achieved with any diagnosis system. Thus, it may be beneficial to carry out a diagnosability analysis before starting the design of a diagnosis system. Structural methods have proven useful for analyzing diagnosability properties of a model and this section is a brief introduction to this topic. The main problem that is described in this section is how a model structure, like the one shown in Figure 2.2, can be utilized to find the faults that are detectable and isolable in the model.

2.3.1. *Fault detectability analysis*

One way to model a fault is to add a new variable f, as in section 2.1, to the model such that $f = 0$ in the fault-free case and $f \neq 0$ in the faulty case. With a slight abuse of notation f will both be referred to as fault signal and its corresponding fault. Without loss of generality, it is possible to assume that a single fault can only violate one equation. If a fault signal f appears in more than one equation, we can replace f in the equations with a new variable x_f and add equation $f = x_f$, which will then be the only equation violated by this fault. Let $e_f \in M$ be the equation that might be violated by a fault $f \in \mathcal{F}$. In the motor model [2.1], faults affect only one equation each and $e_{f_R} = e_1$, $e_{f_{\text{load}}} = e_4$, $e_{f_i} = e_5$, $e_{f_\omega} = e_6$ and $e_{f_T} = e_7$.

A fault f is said to be detectable if there exists an observation that is consistent with fault mode f and inconsistent with the no-fault mode. Let $\mathcal{O}_f$ denote the set of observations consistent with a mode f for a given model M, and let NF denote the fault free mode. Detectability can then formally be defined, as in Nyberg and Frisk (2006):

DEFINITION 2.3.– *A fault mode f is detectable in a model M if $\mathcal{O}_f \setminus \mathcal{O}_{NF} \neq \varnothing$.*

From a structural viewpoint, this means that a detectable fault affects an equation in the structurally overdetermined (SO) part of M, thus structural detectability can be defined as in Blanke *et al.* (2006):

DEFINITION 2.4.– *A fault f is structurally detectable in a model M if $e_f \in M^+$.*

Returning to the running example, the correspondence between detectable faults and structurally detectable faults can be illustrated as follows. The faults f_R, f_i, f_T and f_ω are the detectable faults and residuals capable of detecting them, for example, are

$$\begin{aligned}
&\{e_1, e_5, e_6, e_8\}: \quad r_1 = V - y_i R - L\dot{y}_i - K_f y_i y_\omega = \\
&= y_i f_R - R f_i - L\dot{f}_i - f_i f_R - K_f y_i f_\omega - K_f f_i y_\omega + K_f f_i f_\omega \qquad [2.3] \\
&\{e_3, e_6, e_7, e_9\}: \quad r_2 = J\dot{y}_\omega - y_T + b y_\omega = J\dot{f}_\omega - f_T + b f_\omega.
\end{aligned}$$

The set of equations to the left of the residuals indicates which equations have been used to derive the corresponding residual generator. The expression in only known variables is the computational form of the residual, and the expression in fault variables is the internal form of the residual generator, indicating the fault sensitivity. Thus, fault f_{load} is not detectable.

The Dulmage–Mendelsohn decomposition of the model $M = \{e_1, \ldots, e_9\}$ is shown in Figure 2.5(a), where it can be seen that the SO part of the model is $M^+ = M \setminus \{e_2, e_4\}$. The equations $e_{f_R} = e_1$, $e_{f_i} = e_5$, $e_{f_T} = e_7$ and $e_{f_\omega} = e_6$ corresponding to the detectable faults f_R, f_i, f_T and f_ω belong to M^+, but not the equation $e_{f_{\text{load}}} = e_4$ corresponding to the non-detectable fault f_{load}. This implies, according to definition 2.4, that the detectable faults, f_R, f_i, f_T and f_ω, are the structurally detectable faults in M, which is in accordance with the analytical result above.

2.3.2. *Fault isolability analysis*

A fault f_i is said to be isolable from a fault f_j if there exists an observation that is consistent with fault mode f_i and inconsistent with fault mode f_j and a formal definition is as follows (Nyberg and Frisk 2006).

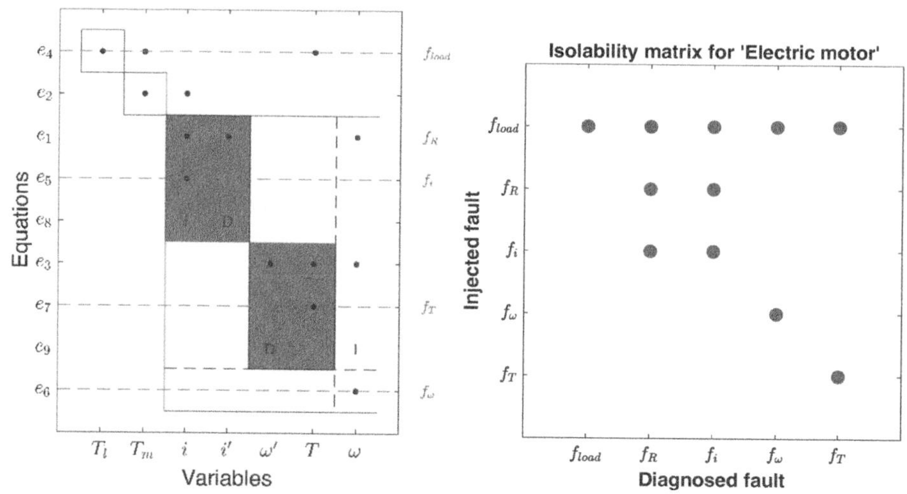

(a) A canonical decomposition. (b) Fault isolability matrix.

Figure 2.5. *Diagnosis analysis results for the model in Figure 2.2. For a color version of this figure, see www.iste.co.uk/puig/diagnosis1.zip*

DEFINITION 2.5.– *A fault mode f_i is isolable from a fault mode f_j in a model M if $\mathcal{O}_{f_i} \setminus \mathcal{O}_{f_j} \neq \varnothing$.*

Detection is a special case of isolation, that is, a fault is detectable if the fault is isolable from the no-fault mode. By noting this similarity, it holds that a fault f_i, isolable from f_j, can violate a monitorable equation in the model describing the behavior of the process having a fault f_j. The set of equations valid with a fault f_j is $M \setminus \{e_{f_j}\}$ and the monitorable part of these equations is, in the generic case, equal to $(M \setminus \{e_{f_j}\})^+$. This motivates the following structural characterization of isolability (Krysander and Frisk 2008; Krysander 2006).

DEFINITION 2.6.– *A fault f_i is structurally isolable from f_j in a model M if*

$$e_{f_i} \in (M \setminus \{e_{f_j}\})^+ \qquad [2.4]$$

To illustrate the definition, consider the case of determining which faults are isolable from sensor fault f_i in the running example. The overdetermined part of $M \setminus \{e_{f_i}\}$, where $e_{f_i} = e_5$, is $\{e_3, e_6, e_7, e_9\}$, that is, the last four equations in Figure 2.5(a). Faults f_T and f_ω belong to these equations, thus only f_T and f_ω are structurally isolable from f_i according to the definition.

Single fault isolability $\mathcal{I}$ for a model with faults F is given by the set of ordered fault pairs (f_i, f_j) such that f_i is structurally isolable from f_j, that is,

$$\mathcal{I} = \{(f_i, f_j) \in F \times F | f_i \text{ is structurally isolable from } f_j\}$$

The faults belonging to $(M \setminus \{e_{f_j}\})^+$ are the faults isolable from f_j. Thus, single fault isolability can be computed as

$$\mathcal{I} = \cup_{f_j \in F} \{(f_i, f_j) | e_{f_i} \in (M \setminus \{e_{f_j}\})^+\}$$

making one Dulmage–Mendelsohn decomposition for each fault in the model.

Single fault isolability $\mathcal{I}$ can be visualized with a so-called fault isolability matrix (Krysander 2006). The isolability matrix for the running example is shown in Figure 2.5(b). Both rows and columns correspond to the faults, and a dot in position (f_i, f_j) means that f_i is not isolable from f_j, that is, $(f_i, f_j) \notin \mathcal{I}$. Unique isolation of all faults corresponds to the identity matrix. The dots in row f_i correspond to the diagnosed faults under the condition that fault f_i is the injected fault. The part of the fault isolability matrix corresponding to detectable faults is symmetric, and in this example the detectable faults are f_R, f_i, f_ω and f_T. The interpretation of the isolability matrix in Figure 2.5(b) is that the fault f_{load} is not isolable from any other faults because this fault is not even detectable. The block corresponding to f_R and f_i indicates that these faults are not isolable from each other but isolable from all other faults. Finally, f_ω and f_T are uniquely isolable.

2.3.3. *Canonical isolability decomposition of the overdetermined part*

The Dulmage–Mendelsohn decomposition reveals the overdetermined part of a model. The overdetermined part can be decomposed into a finer decomposition based on isolability properties of the model and this finer decomposition is described in the following.

Consider a model where the overdetermined part of the model is equal to the model itself, that is, $M = M^+$. Such a model can be partitioned into a number of equation sets M_k, as shown in Figure 2.6, such that faults f_i and f_j are structurally isolable if and only if the faults belong to different equation sets M_i. The white parts of the matrix have only zero entries. The partitioning consists of m equation sets M_i in the figure; the n first equation sets M_i correspond to the n blocks (M_i, X_i), for $i \in \{1, 2, \ldots, n\}$ in the matrix, and the $m - n + 1$ last equation sets M_i contain only one equation each and are not related to any variables.

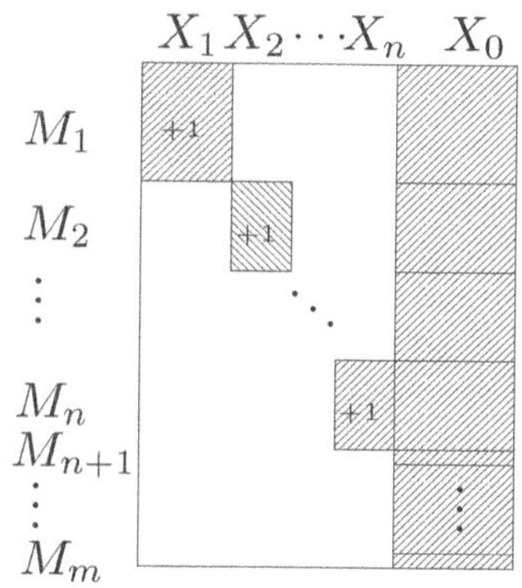

Figure 2.6. *Canonical decomposition of the overdetermined part of a model*

This partitioning can be computed as follows. Given an equation $e \in M$, the equation set M_i including equation e can be computed as

$$M_i = M \setminus (M \setminus \{e\})^+ \qquad [2.5]$$

To understand the relation between this decomposition and structural isolability, we need to note the similarity between [2.5] and definition 2.6.

Let e in [2.5] be e_f. The faults in $(M \setminus \{e_f\})^+$ are the faults isolable from f according to definition 2.6, which means that the faults in the complement set M_i contain all faults not isolable from f. Thus, faults in the same set are not isolable from one another and faults in different sets are isolable from each other. More details of the decomposition is given in Krysander and Frisk (2008).

Consider the running example again and the decomposition in Figure 2.5(a). First, note that $\{e_2, e_4\}$ is the exactly determined part and $\{e_1, e_3, e_5, e_6, e_7, e_8, e_9\}$ is the overdetermined part. The finer decomposition of the overdetermined part consists of the blocks $\{e_1, e_5, e_8\}$, $\{e_3, e_7, e_9\}$ and $\{e_6\}$ with faults $\{f_R, f_i\}$, $\{f_T\}$ and $\{f_\omega\}$, respectively. The fault sets correspond to the diagonal blocks in the isolability matrix in Figure 2.5(b). The faults $\{f_R, f_i\}$ are not isolable from each other because they are in the same block and the reason for this is that there are too few equations $\{e_1, e_5, e_8\}$ relating variables i and i'. The faults f_T and f_ω have their own blocks, thus they are uniquely isolable. This is of course in agreement with Figure 2.5(b).

Single fault isolability is summarized in the fault isolability matrix in Figure 2.5(b), as previously mentioned. In this analysis, dynamics is not considered, that is, the additional information that there are D:s and I:s in the structural model, but there are extensions that takes this into account, see Frisk *et al.* (2012) and

Frisk *et al.* (2017). For example, standard observer-based techniques solve differential equations by integration only, that is, with integral causality. Under this restriction of causality, the structural isolability might be reduced, but for the electric motor the same isolability is obtained under integral causality.

2.4. Testable submodels

Residual generation using structural methods was first introduced by M. Staroswiecki and described in Blanke *et al.* (2006). Many methods have since been described, see Armengol *et al.* (2009) for a comparative study. What structural analysis allows is the determination of the set of constraints or equations, that is, testable submodels from which a residual can be generated and, furthermore, the analysis can also provide a computational sequence. The main objective of this section is to present algorithms to find all the testable subsystems and show how they can be used to generate residuals. Again, note that the results are structural which means that they should be considered generic or best-case results but are still a good starting point for a well-tuned diagnosis system.

2.4.1. *Basic definitions*

From the electric motor model [2.1], the residuals given in [2.3] have been derived by algebraic elimination of the known variables. Both equations are called analytical redundancy, relations or parity relations, and can be used to check if u, y_1 and y_2 are consistent with the model [2.1]. This means that the equation sets given in [2.3] are testable submodels. Also, this example shows the fault influences on the residuals determining detectability and isolability capability of the residuals. These sets of faults are called *test supports* (TSs) (Krysander *et al.* 2010).

Following definition 2.4, the SO part corresponds to the testable subset, that is, $M^+ \neq \varnothing$ is a testable submodel. This motivates the following structural characterization of testable submodels (Krysander *et al.* 2008).

DEFINITION 2.7.– *A non-empty set of equations, $M \neq \varnothing$, is a proper structurally overdetermined (PSO) set if $M = M^+$.*

Further, it is of interest to find the smallest PSO sets, since the more local the testable subset is, the more fault isolation information is obtained when the corresponding residual generates an alarm.

DEFINITION 2.8.– *A PSO set is a minimal structurally overdetermined (MSO) set if no proper subset is a PSO set.*

For linear systems, the dimension of the residual space quantifies the degree of redundancy of a model and there is a corresponding structural property.

DEFINITION 2.9.– *The degree of structural redundancy of a model M is defined as*

$$\varphi(M) = |M^+| - |\mathcal{X}^+|$$

EXAMPLE 2.1.– To illustrate these definitions, consider the model in [2.1], its corresponding structural model in Figure 2.2 and Dulmage–Mendelsohn decomposition in Figure 2.5(a). Then,

– $M^- = \varnothing$, $M^0 = \{e_2, e_4\}$, and $M^+ = \{e_1, e_3, e_5, e_6, e_7, e_8, e_9\}$.

– M^+ is a PSO with a redundancy

$$\varphi(M) = |M^+| - |\mathcal{X}^+| = 7 - 5 = 2 > 0$$

– M^+ is not an MSO set because there are PSO subsets. In this example, there are three MSO sets

$$\begin{aligned} &MSO_1 : \{e_3, e_6, e_7, e_9\} \text{ with } \varphi(MSO_1) = 4 - 3 = 1 \\ &MSO_2 : \{e_1, e_5, e_6, e_8\} \text{ with } \varphi(MSO_2) = 4 - 3 = 1 \\ &MSO_3 : \{e_1, e_3, e_5, e_7, e_8, e_9\} \text{ with } \varphi(MSO_3) = 6 - 5 = 1 \end{aligned} \quad [2.6]$$

Note that an MSO set is a PSO set with a structural degree of redundancy of 1. We can also conclude that each testable submodel has a corresponding MSO model. MSO and PSO models characterize model redundancy but faults are not taken into account. The residuals presented in [2.3] have two parts, the computational form and the internal form describing the fault influence. As shown below, the three MSOs are characterized by a set of equations and are also sensitive to a set of faults.

$$\begin{aligned} &\text{Faults}(MSO_1) = \{f_\omega, f_T\} \\ &\text{Faults}(MSO_2) = \{f_R, f_i, f_\omega\} \\ &\text{Faults}(MSO_3) = \{f_R, f_i, f_T\} \end{aligned} \quad [2.7]$$

Now, let $F(M)$ denote the set of faults that influence any of the equations in M. Then, since a PSO set exactly characterizes a set of equations that can be used to form a test, a formal definition is then given by Krysander *et al.* (2010):

DEFINITION 2.10.– *Given a model $\mathcal{M}$ and a set of faults $\mathcal{F}$, a non-empty subset of faults $\zeta \subseteq \mathcal{F}$ is a test support if there exists a PSO set $M \subseteq \mathcal{M}$ such that $F(M) = \zeta$.*

It is also important to characterize sets of equations that can be used to form a test.

DEFINITION 2.11.– *An equation set M is a test equation support (TES) if*

1) $F(M) \neq \emptyset$;

2) M is a PSO set;

3) for any $M' \not\supseteq M$ where M' is a PSO set it holds that $F(M') \not\supseteq F(M)$.

Here, it is interesting to consider minimal test support (MTS) and minimal TES (MTES).

DEFINITION 2.12.– *Given a model, a test support is an MTS if no proper subset is a test support.*

DEFINITION 2.13.– *A TES M is an MTES if there exists no subset of M that is a TES.*

In [2.6], all MSO sets are also MTES but this is not true in general. See Krysander *et al.* (2010) for more details.

2.4.2. *MSO algorithm*

An efficient algorithm for computing the set of all MSO sets is presented in Krysander *et al.* (2010). It is based on examining the set M of constraints of a proper structurally over-constrained structure graph. The proposed function is based on a top-down approach in the sense that we start with the entire model and then reduce the size of the model step by step until an MSO set remains. The basic version of an efficient algorithm is straightforward, and the function `FindMSO` computes the set of all MSO sets contained in the model. Assume, without loss of generality, that the input model fulfills $M = M^+$.

```
function 𝓜_MSO = FindMSO(M)
    if φ(M) = 1
        𝓜_MSO := {𝓜}
    else
        𝓜_MSO := ∅
        for each  e ∈ M
            M' := (M \ {e})^+
            𝓜_MSO := 𝓜_MSO ∪ FindMSO(M')
        end
    end
end
```

To illustrate the steps in this function, consider the electric motor model [2.1] with the corresponding structural representation in Figure 2.4(a).

The structural redundancy of $M = \{e_1, \ldots, e_9\}$ is $\varphi(M) = 2$ and the for-loop is entered in the function `FindMSO`. First e_1 is removed and the overdetermined part of the new set $M \setminus \{e_1\} = \{e_3, e_5, e_6, e_7, e_8, e_9\}$ is $M_1 = (M \setminus \{e_1\})^+ = \{e_3, e_6, e_7, e_9\}$. The set M' in the algorithm becomes $(M \setminus \{e_1\})^+ = \{e_3, e_6, e_7, e_9\}$. Now, $\varphi(M') = 1$ and M_1 is saved as an MSO in $\mathcal{M}_{MSO}$. Then e_3 is removed, and the MSO set $M_2 = (M \setminus \{e_3\})^+ = \{e_1, e_5, e_6, e_8\}$ is found, which also has redundancy 1 and is an MSO. Next, e_5 is removed and $(M \setminus \{e_5\})^+ = \{e_3, e_6, e_7, e_9\}$, that is, an MSO equal to M_1 already identified above. Next, e_6 is removed and $M_3 = (M \setminus \{e_6\})^+ = \{e_1, e_3, e_5, e_7, e_8, e_9\}$, which has redundancy 1, that is, an MSO set. Continuing like this will show that these are the only three MSO sets.

It has been noted that, in the approach above, the same MSO set can be found more than once, which decreases efficiency. To avoid this situation, and guarantee that each MSO set is only found once, equivalence classes defined on the set of equations are useful. Let R be a relation, defined on the set of equations M, defined by $(e', e) \in R$ if

$$e' \notin (M \setminus \{e\})^+, \qquad [2.8]$$

this implies that $(M \setminus \{e\})^+ = (M \setminus \{e'\})^+$, that is, removing e and e' and then extracting the overdetermined part results in the same set of equations. Since removing an equation and then computing the overdetermined part is a key step in the function `FindMSO` (line 7), equivalence classes can be used to avoid finding the same MSO set more than once. The set M can be partitioned into m equivalence classes M_i. The equivalence classes can then be lumped together forming a reduced structure. Full details of this algorithm are described in Krysander *et al.* (2008). Note also that this partitioning of equations is the same as that introduced in section 2.3.3.

2.4.3. *Residual generation based on matching*

Each MSO set comprises a set of constraints where at least one constraint can be used as an *analytical redundancy relation* (ARR), defined as follows.

DEFINITION 2.14.– *Let M be a model and $\mathcal{O}$ the set of observations, then an equation $r(\mathcal{O}) = 0$ is an ARR for $M(\mathcal{O})$ if for each $o_i \in \mathcal{O}$ consistent with the model $M(\mathcal{O})$, the equation is fulfilled.*

One way to obtain an ARR is to find a complete matching of the unknown variables in an MSO set, use that matching to compute all unknown variables, and then compute a residual using the redundant constraint.

Choosing a complete matching on unknown variables of the MSOs make it possible to rearrange the equations of M and to rewrite them for designing the residual generator.

DEFINITION 2.15.– *A model taking a subset of the variables* $\boldsymbol{o} \subset \mathcal{O}$ *as input, and generating a scalar signal* r *as output, is a residual generator for the model* $M(\boldsymbol{o}, \boldsymbol{x})$, *if for all* $\boldsymbol{o}$ *consistent with* $M(\boldsymbol{o}, \boldsymbol{x})$, *it holds that* $\lim_{t \to \inf} = 0$.

Note that for an MSO set, if all constraints are invertible with respect to all variables, any constraint in the MSO can be used as a residual equation. This does not mean that they all have equivalent performance or properties, which is illustrated next. But in general, not all the potential MSOs make it possible to generate residuals and not all matchings can lead to a residual. This matter is linked with the conditions under which a variable substitution can actually be performed. A given relation can be interpreted in a causal way, and the substitution of variables in one ARR can be performed when having consistent causalities. Classical non-invertible constraints are nonlinear functions and conditional functions, and we can also include the differential constraints in this group.

To illustrate, consider the MSO set $M_3 = \{e_1, e_3, e_5, e_7, e_8, e_9\}$ for the electric motor example. This set has six constraints in five variables and any constraint can be used as a redundant/residual equation. Figure 2.7 shows computational graphs when choosing e_5, e_1 and e_3 respectively, as a redundant equation. This leads to residual generators in integral, mixed and differential causality, respectively. To see this, in Figure 2.7(a), there is only computational flow *from* a differentiated variable *to* a non-differentiated variable, e.g. from ω' to ω (i.e. integration). In Figure 2.7(c), the case is opposite, e.g. from i to i' (i.e. differentiation). And finally, in Figure 2.7(b), both cases are represented, e.g. integration of ω' to ω and differentiation of i to i'.

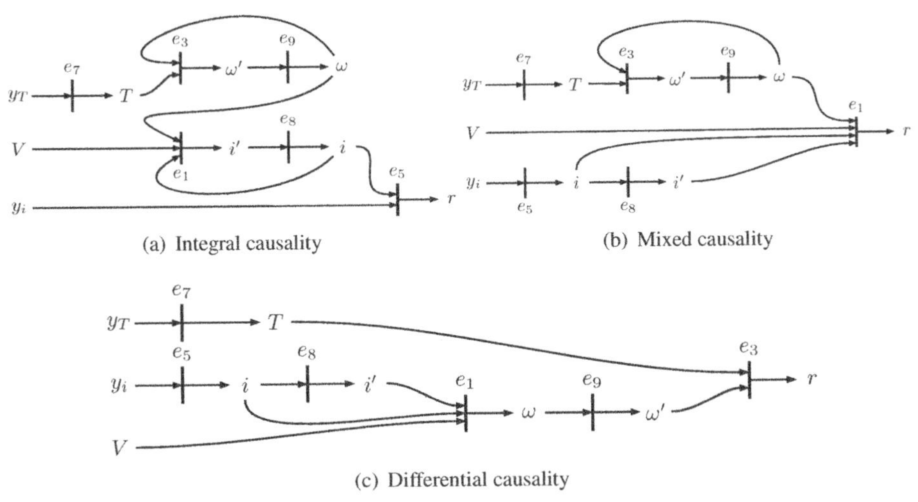

Figure 2.7. *Computational graphs for MSO set* M_3 *with different residual equations*

2.5. Sensor placement

The number and location of sensors significantly influence the expected fault detection and isolation performance. This section describes an approach for computing a characterization of all possible sets of sensors that meets the requirements, from a given model and specified detectability and isolability performance specification. The main ideas and results are presented; for full details, proofs, optimizations, and further properties, see Krysander and Frisk (2008).

2.5.1. *The basic sensor placement problem*

To illustrate the fundamental problem, consider a fifth-order linear model

$$\begin{aligned} e_1 : \quad & \dot{x}_1 = -x_1 + x_2 + x_5 \\ e_2 : \quad & \dot{x}_2 = -2x_2 + x_3 + x_4 \\ e_3 : \quad & \dot{x}_3 = -3x_3 + x_5 + f_1 + f_2 \\ e_4 : \quad & \dot{x}_4 = -4x_4 + x_5 + f_3 \\ e_5 : \quad & \dot{x}_5 = -5x_5 + u + f_4 \end{aligned} \qquad [2.9]$$

where x_i are the state variables, u is a known control signal and f_i are the faults we want to detect and isolate. Since there are no sensors, there is no redundancy and the faults are not detectable. Then the question is: Which sensors should be added to detect and isolate the faults? Before answering this, let us introduce some notation and let F denote the set of faults. A detectability performance specification is then a set $F_{det} \subseteq F$ specifying the detectability requirement and an isolability requirement is a set I of ordered pairs $(f_i, f_j) \in F_{det} \times F_{det}$, meaning that f_i is isolable from f_j. Note that we assume that all faults that are included in the isolability specification I are also required to be detectable.

Since the fault isolability capability always increases when adding new sensors (for the moment, consider new sensors to be fault free), there are minimal elements in the family of sensor sets that achieve a certain level of fault isolability. Therefore, a *minimal sensor set* is defined as follows.

DEFINITION 2.16 (Minimal sensor set).– *Let $\mathcal{S}$ be the set of possible sensor locations, that is, the set of measurable variables, and let S be a multiset defined on $\mathcal{S}$. Then S is a minimal sensor set, with respect to a given detectability and isolability specification, if adding the sensors in S fulfills the specification and all proper subsets of S do not.*

Note that S is a multiset, which is similar to a set but allows for multiple instances of a member. Multisets are used as it may be necessary to add more than one sensor measuring the same variable if the new sensors can fail.

Going back to the small linear model, it can be verified, using methods described in section 2.3, that there are five minimal sensor sets that achieve maximal fault isolation: $\{x_1, x_3\}$, $\{x_1, x_4\}$, $\{x_2, x_3\}$, $\{x_2, x_4\}$ and $\{x_3, x_4\}$. Thus, adding sensors measuring the variables in any of these sets, or a superset of the variables, achieves maximum fault isolability. Now, it could be the case that the new sensors may also become faulty. If maximum fault isolability is also desired for faults in the new sensors, there are nine minimal sensor sets where one sensor set is two sensors measuring x_1 and one for x_3, that is, the multiset $S = \{x_1, x_1, x_3\}$ is a minimal sensor set.

2.5.2. *A structural approach*

This section describes how to compute all minimal sensor sets, and a general assumption here is that the model does not contain any underdetermined part. This means, for example, that the model can be simulated with a unique solution given an initial state. Without loss of generality, it is also assumed that (1) no fault affects more than one equation and (2) possible sensors measure a function of one unknown variable. Both these assumptions can be fulfilled by introducing additional variables. For example, consider the case where a sensor measures some function $h(x)$ rather than just a single variable. Then, we introduce a new variable x_{new}, an equation $x_{new} = h(x)$ and then the new measurement equation will then be $y = x_{new}$. The approach consists of three steps: (1) how to select sensors to achieve detectability, (2) how to rewrite isolability requirements as detectability requirements and then (3) put it all together.

2.5.2.1. *Sensor placement for detectability*

Consider again the small example model and fault f_3. To make this fault detectable, according to definition 2.4, an additional sensor is needed such that equation $e_{f_3} = e_4$ is in the overdetermined part of the model. It is straightforward to verify that f_3 becomes detectable if and only if any one of the variables $\{x_1, x_2, x_4\}$ is measured. To understand why exactly these variables give detectability of f_3, consider first Figure 2.8(a) where the model structure is shown. Here, the model structure is already in a triangular form, but in general the Dulmage–Mendelsohn decomposition has to be carried out first to get the upper triangular form.

The strong components are denoted as b_i and, for example, block b_1 is said to be connected with b_2 via a non-zero element in position $(1, 2)$ and in a similar fashion b_2 is connected to b_4. Such connections in the structure graph then imply an *ordering* among the strong components, e.g. $b_2 \leq b_1$ and $b_5 \leq b_2$. This ordering is a partial order, and the Hasse diagram is shown in Figure 2.8(b). The interpretation of the Hasse diagram is that if there is an edge *from* b_j to b_i, this means that $b_j \leq b_i$. Note that it is a partial order, that is, not all strong components are ordered. For example, there is no order between b_3 and b_4. It can be shown that this ordering makes it possible to state exactly which equations, in an exactly determined model, become overdetermined when adding a sensor. Each block b_i is directly related to an equation set M_i, given by the Dulmage–Mendelsohn decomposition.

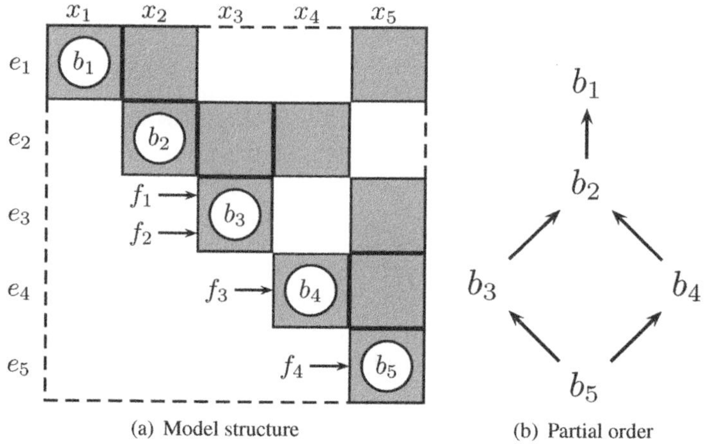

Figure 2.8. *Model structure and Hasse diagram of the block partial order over the strongly connected components for the small example model*

THEOREM 2.1 (Krysander and Frisk 2008).– Let M be an exactly determined set of equations, b_i a strongly connected component in M with equations M_i and $e \notin M$ an equation corresponding to measuring any variable in b_i. Then

$$(M \cup \{e\})^+ = \{e\} \cup (\cup_{b_j \leq b_i} M_j) \qquad [2.10]$$

It follows from theorem 2.1 that measuring a variable in a block ordered higher than the block the fault enters achieves detectability. To use this result for sensor selection, let $P \subseteq X$ be a set of possible sensor locations and introduce the set

$$D(f_i) = \{x | b_i \leq b_j, x \in X_j \cap P\} \qquad [2.11]$$

where X_j is the set of variables corresponding to block b_j and b_i is the block that is influenced by the fault f_i. The set $D(f_i)$ is then the set of variables such that measuring *any* variable in the set achieves detectability of f_i. Consider again the fault f_3 in the small example that affects an equation in b_4 in Figure 2.8. To make f_3 detectable, *any* variable in b_1, b_2 or b_4 should be measured, that is, $D(f_3) = \{x_1, x_2, x_4\}$.

Further, this means that to find sensors that achieve detectability of a *set* of faults, compute detectability sets $D(f_i)$ for each fault f_i and select a sensor set that has a non-empty intersection with each $D(f_i)$. This means that a solution measures at least one variable that makes each fault detectable. Thus, finding the sensor set corresponds to finding a *hitting set* for the sets $D(f_i)$. To obtain minimal solutions, a minimal hitting set algorithm (Reiter 1987; de Kleer and Williams 1987) can be directly applied to find all minimal sensor sets that achieve detectability. For the example model, the corresponding detectability sets for all faults are

$$D(f_1) = D(f_2) = \{x_1, x_2, x_3\}, \quad D(f_3) = \{x_1, x_2, x_4\}, \quad D(f_4) = X$$

and minimal sensor sets that achieve detectability of all faults are then $\{x_1\}$, $\{x_2\}$ and $\{x_3, x_4\}$, which are the same sensor sets as in section 2.5.1.

2.5.2.2. *Sensor placement for isolability of detectable faults*

This section describes how to find sensors such that detectable faults are also isolated. Achieving maximum isolability of a set of single faults F can be divided into $|F|$ sub-problems, one for each fault, as follows. For each fault $f_j \in F$, find all measurements that make the maximum possible number of faults $f_i \in F \setminus \{f_j\}$ isolable from f_j. The solution to the isolability problem will then be obtained by combining the results from all sub-problems.

Now, to show how this can be done, each sub-problem will be formulated as a detectability problem. Assume that M is a model, including sensors such that all faults are detectable, and M_S represents a set of equations describing an additional sensor set S, e.g. if one sensor is added measuring a variable x, then M_S consists of the equation $y = x$. Let $M' = M \setminus \{e_{f_j}\}$, then given the sensor set S, a fault f_i is isolable from f_j in the model $M \cup M_S$ according to definition 2.6 if

$$e_{f_i} \in (M' \cup M_S)^+. \qquad [2.12]$$

Then, according to definition 2.4, this means that f_i is structurally detectable in $M' \cup M_S$. Hence, the maximum possible number of faults $f_i \in F \setminus \{f_j\}$ is *isolable* from f_j in $M \cup M_S$ if the maximum possible number of faults $f_i \in F \setminus \{f_j\}$ is structurally *detectable* in the model $(M \setminus \{e_{f_j}\}) \cup M_S$. Thus, each isolability sub-problem can be formulated as a detectability problem.

To outline the solution of one such sub-problem, again consider the small linear example. Assume that we have added sensors measuring $\{x_3, x_4\}$ such that all faults are detectable. Furthermore, we assume that these sensors can be faulty and denote these new sensor faults as f_5 and f_6, respectively. A row permuted structure of the obtained model $M = \{e_1, e_2, \ldots, e_7\}$ is shown in Figure 2.9.

Consider now the sub-problem associated with fault f_1, that is, isolate as many faults as possible from f_1. The set M' in [2.12] is equal to $M' = M \setminus \{e_{f_1}\} =$

$M \setminus \{e_3\}$. The sub-problem is then, given the model M', to find the minimal additional sensors S such that as many of the faults $f_2, f_3, \ldots, f_6$ as possible become detectable in $M' \cup M_S$.

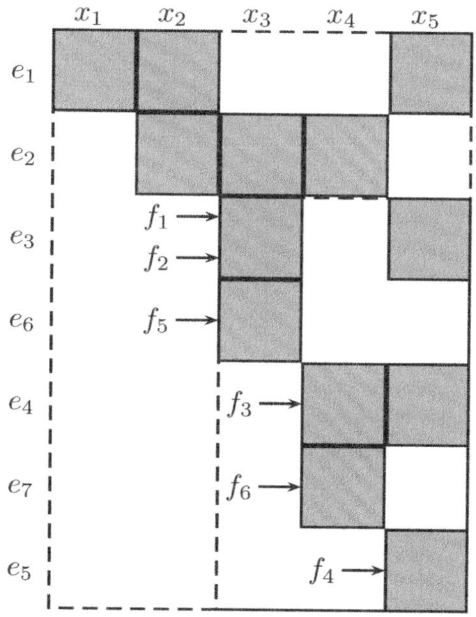

Figure 2.9. *Block structure of the example in section 2.5.1 extended with measurements of x_3 and x_4*

The faults can, depending on which equations they violate, be divided into the following three types: faults that do not violate any equation in M', faults that violate equations in the SO part $(M')^+$ and faults that violate other equations in M', that is, $M' \setminus (M')^+$. In the example, we have that $(M')^+ = \{e_4, e_5, e_7\}$ and $M' \setminus (M')^+ = \{e_1, e_2, e_6\}$, which is equal to the structurally just-determined part of M'. This implies that f_2 is not included in M'; f_3, f_4 and f_6 belong to the SO part, and f_5 belongs to the structurally just-determined part. Fault f_2 is not included in M' and cannot be structurally detectable in $M' \cup M_S$ for any sensor set S. This implies that f_2 is not isolable from f_1 with any sensor addition and this also follows from the fact that these two faults violate the same equation. The faults f_3, f_4 and f_6 in the SO part $(M')^+$ are, according to definition 2.4, structurally detectable in M' and require no additional measurements. Fault f_5 in the just-determined part is not detectable, but f_5 can become detectable in $M' \cup M_S$ if S is appropriately selected.

By applying the approach from section 2.5.2.1 to the structurally just-determined part of M', that is, the subgraph of M' defined by the node sets $\{e_1, e_2, e_6\}$ and

$\{x_1, x_2, x_3\}$, we get that $D(f_5) = \{x_1, x_2, x_3\}$. Hence, one of the variables in the detectability set $\{x_1, x_2, x_3\}$ must be measured to make the faults $F \setminus \{f_1, f_2\}$ detectable in $M' \cup M_S$ and this implies that all faults in $F \setminus \{f_1, f_2\}$ are isolable from f_1 in $M \cup M_S$. The solution to the sub-problem related to fault f_1 will be the computed detectability set. The next result formalizes the solution of a sub-problem.

THEOREM 2.2.– Let M be a set of equations with no structurally underdetermined part, F a set of structurally detectable faults in M, $P \subseteq X$ the set of possible sensor locations and M_S the equations added by adding the sensor set S.

For an arbitrary fault f_j, let M^0 be the just-determined part of $M \setminus \{e_{f_j}\}$, F^0 the set of faults contained in M^0 and $\mathcal{D}$ detectability sets for faults F^0 in M^0 according to the procedure in section 2.5.2.1.

Then, the maximum possible number of faults $f_i \in F \setminus \{f_j\}$ is structurally isolable from f_j in $M \cup M_S$ if and only if S has a non-empty intersection with all non-empty sets in $\mathcal{D}$.

For the example shown in Figure 2.9, the families of detectability sets of the different sub-problems are

$$\begin{aligned} &\{\{x_1, x_2, x_3\}\} \text{ for } f_1, f_2 \text{ and } f_5 \\ &\{\{x_1, x_2, x_4\}\} \text{ for } f_3 \text{ and } f_6 \\ &\varnothing \text{ for } f_4 \end{aligned} \qquad [2.13]$$

We have found two distinct detectability sets and the minimal hitting sets are $\{x_1\}$, $\{x_2\}$ and $\{x_3, x_4\}$. These sets are the minimal additional measurements that achieve maximum single fault isolability.

2.5.2.3. *Sensor placement for both detectability and isolability*

The discussion above shows how isolability can be achieved in a model where all faults are structurally detectable. Next, the approach will be extended to models where faults may not be structurally detectable in the original model. Again, the solution is first outlined for the small linear example.

The faults in the model, without any sensors, are not detectable and the objective is to find all minimal sensor sets that maximize fault detectability and isolability. It has been previously shown that the minimal sets of measurements to achieve full detectability are $\{x_1\}$, $\{x_2\}$ and $\{x_3, x_4\}$.

If we start by adding the first set, that is, a sensor measuring x_1 described by an equation e_s, a new model $M \cup \{e_s\}$ is obtained where all faults are detectable. Since all faults are detectable, the previously described method to achieve maximum isolability can be applied to the model $M \cup \{e_s\}$. The minimal sensor sets that solve this problem are $\{x_3\}$ and $\{x_4\}$. By combining this result with the fact that a sensor measuring x_1 has been added to obtain detectability, it follows that $\{x_1, x_3\}$ and $\{x_1, x_4\}$ are two possible sensor sets that achieve maximum detectability and isolability. To compute all minimal sensor sets that achieve maximum isolability, we also have to investigate the solutions when we decide to measure $\{x_2\}$ or $\{x_3, x_4\}$ to obtain full detectability. By solving one isolability problem for each of the minimal sensor sets that achieves full detectability, we get that the minimal sensor sets are $\{x_1, x_3\}$, $\{x_1, x_4\}$, $\{x_2, x_3\}$, $\{x_2, x_4\}$ and $\{x_3, x_4\}$, which are the same sets as in section 2.5.1.

The above procedure for the small linear example can be directly extended to any model. All operations needed have polynomial complexity except for the minimal hitting set algorithm which is NP-hard. This means that the worst case might be intractable, however this is a direct consequence of the objective to find *all* minimal sensor sets. If the objective is to find not all, but small sensor sets, approximate hitting set algorithms (de Kleer 2011) can be applied, which makes large problems tractable. Note, however, that well-formed models of physical systems typically have a structure, which makes the computations less demanding. Also, and this is possibly the most important aspect, the number of measurable signals in the specification P is a prime indicator of the complexity. Thus, it is not primarily the number of equations or number of faults, but rather the user-specified sensor specification that controls the complexity.

2.6. Summary and discussion

In this chapter we have introduced structural analysis for model analysis and diagnosis system design. We have introduced the properties of fault detectability and isolability of a model which both make checking the limit on diagnosis performance possible. The model structure has also been used for computing all testable equation sets from which analytical redundancy relations or sequential residual generators can be designed. Computation sequences for the sequential residual generators have been provided.

One important step in structurally aided diagnosis system design not covered in this chapter is test selection. The number of testable equation sets is exponential in model redundancy and often a small number of tests are sufficient for satisfactory diagnosis performance. Further reading on test selection methods are, for example, Krysander *et al.* (2010) where residual selection is based on which equation each fault affects, Frisk *et al.* (2017) where the so-called differential index indicating the difficulty of simulating residuals is considered, and Frisk and Krysander (2018) and Jung and Frisk (2018) where machine learning methods use no-fault and faulty measurement data for selecting a small set of tests with good diagnosability performance. In the latter approaches model uncertainties and measurement noise are considered through the use of measurement data. These selection strategies can be used individually or in combination.

The number and location of sensors significantly influences the expected fault detection and isolation performance. As a complement to the methods presented here, it is worth mentioning that it is possible to compute where to put sensors in order to achieve a specified detectability and isolability performance. There are methods that compute a characterization of all possible sets of sensors that meet the requirements given a structural model of the system. The main ideas can be found in Frisk *et al.* (2019), for full details, proofs, optimizations and further properties, see Krysander and Frisk (2008).

2.7. References

Armengol, J., Bregón, A., Escobet, T., Gelso, E., Krysander, M., Nyberg, M., Olive, X., Pulido, B., Travé-Massuyès, L. (2009). Minimal structurally overdetermined sets for residual generation: A comparison of alternative approaches. *IFAC Proceedings Volumes*, 42(8), 1480–1485.

Blanke, M., Kinnaert, M., Lunze, J., Staroswiecki, M. (2006). *Diagnosis and Fault-Tolerant Control*. Springer-Verlag, Berlin, Heidelberg.

Cassar, J.P. and Staroswiecki, M. (1997). A structural approach for the design of failure detection and identification systems. *IFAC Proceedings Volumes*, 30(6), 841–846.

de Kleer, J. (2011). Hitting set algorithms for model-based diagnosis. *Proceedings of the 22nd International Workshop on Principles of Diagnosis (DX-11)*, October 4–7. Murnau, Germany.

de Kleer, J. and Williams, B.C. (1987). Diagnosing multiple faults. *Artificial Intelligence*, 32, 97–130.

Dulmage, A.L. and Mendelsohn, N.S. (1958). Coverings of bipartite graphs. *Canadian Journal of Mathematics*, 10, 517–534.

Frisk, E., Bregon, A., Åslund, J., Krysander, M., Pulido, B., Biswas, G. (2012). Diagnosability analysis considering causal interpretations for differential constraints. *IEEE Transactions on Systems, Man, and Cybernetics – Part A: Systems and Humans*, 42(5), 1216–1229.

Frisk, E. and Krysander, M. (2018). Residual selection for consistency based diagnosis using machine learning models. *IFAC Symposium on Fault Detection, Supervision and Safety for Technical Processes (SAFEPROCESS)*, Warszaw, Poland.

Frisk, E., Krysander, M., Åslund, J. (2017). Analysis and design of diagnosis systems based on the structural differential index. *IFAC World Congress*, Toulouse, France.

Frisk, E., Krysander, M., Jung, D. (2017). A toolbox for analysis and design of model based diagnosis systems for large scale models. *IFAC-PapersOnLine*, 50(1), 3287–3293.

Frisk, E, Krysander, M, Escobet, T. (2019). Structural analysis. In *Fault Diagnosis of Dynamic Systems*, Escobet, T., Bregon, A, Pulido, B., Puig, V. (eds). Springer, Cham, 43–68. Available at: https://doi.org/10.1007/978-3-030-17728-7_3.

Jung, D. and Frisk, E. (2018). Residual selection for fault detection and isolation using convex optimization. *Automatica*, 97, 143–149.

Krysander, M. (2006). Design and analysis of diagnosis systems using structural methods. PhD Thesis, Linköping University, Linköping, Sweden.

Krysander, M., Åslund, J., Frisk, E. (2010). A structural algorithm for finding testable sub-models and multiple fault isolability analysis. *21st International Workshop on Principles of Diagnosis (DX-10)*, Portland, USA.

Krysander, M., Åslund, J., Nyberg, M. (2008). An efficient algorithm for finding minimal overconstrained subsystems for model-based diagnosis. *IEEE Transactions on Systems, Man, and Cybernetics – Part A: Systems and Humans*, 38(1), 197–206.

Krysander, M. and Frisk, E. (2008). Sensor placement for fault diagnosis. *IEEE Transactions on Systems, Man, and Cybernetics – Part A: Systems and Humans*, 38(6), 1398–1410.

Murota, K. (2009). *Matrices and Matroids for Systems Analysis*. Springer, Berlin, Heidelberg.

Nyberg, M. and Frisk, E. (2006). Residual generation for fault diagnosis of systems described by linear differential-algebraic equations. *IEEE Transactions on Automatic Control*, 51(12), 1995–2000.

Ploix, S., Désinde, M., Touaf, S. (2005). Automatic design of detection tests in complex dynamic systems. *IFAC Proceedings Volumes*, 38(1), 478–483.

Pulido, B. and González, C.A. (2004). Possible conflicts: A compilation technique for consistency-based diagnosis. *IEEE Transactions on Systems, Man, and Cybernetics – Part B: Cybernetics*, 34(5), 2192–2206.

Reiter, R. (1987). A theory of diagnosis from first principles. *Artificial Intelligence*, 32(1), 57–95.

3

Set-based Fault Detection and Isolation

Ye WANG[1] and Vicenç PUIG[2]

[1] *Department of Electrical and Electronic Engineering, University of Melbourne, Australia*

[2] *Institute for Robotics and Industrial Informatics, Polytechnic University of Catalonia, Barcelona, Spain*

3.1. Introduction

Model-based fault detection of dynamic processes is based on the use of models (i.e. *analytical redundancy*) to check the consistency of observed behaviors. However, when building a model of a dynamic process to monitor its behavior, there is always some mismatch between the modeled and real behavior, due to the fact that some effects are neglected, some nonlinearities are linearized in order to simplify the model, some parameters have tolerance when they are compared between several units of the same component, some errors in parameters (or in the structure) of the model are introduced in the model identification process, etc. These modeling errors introduce some uncertainty into the model. Usually, this uncertainty can be bounded and included in the fault detection model. There are several ways of considering the uncertainty associated with the model, depending if is located in the parameters (*structured*) or in the model structure (*non-structured*). In the literature, a fault diagnosis algorithm that is able to handle uncertainty is called *robust*. The *robustness* of a fault diagnosis algorithm is the degree of sensitivity to faults compared to the degree of sensitivity to uncertainty (Chen and Patton 1999). Research on robust fault diagnosis methods has been very active in the fault detection and isolation (FDI) community these past years. One of the most developed families of approaches,

Diagnosis and Fault-tolerant Control 1,
coordinated by Vicenç PUIG and Silvio SIMANI.

known as *active*, is based on generating residuals that are insensitive to uncertainty, while at the same time sensitive to faults. This approach has been extensively developed by several researchers using different techniques: unknown input observers (UIO), robust parity equations, H_∞, etc. In Chen and Patton (1999), there is an excellent survey of this active approach. On the other hand, there is a second family of approaches, known as *passive*, which enhances the robustness of the fault detection system at the decision-making stage by propagating the uncertainty to the residuals and generating an *adaptive threshold*. Seminal papers suggesting this approach are Horak (1988) in the time domain and Emami-Naeini *et al.* (1988) in the frequency domain. This passive approach has been developed in recent years by several studies but still is under research; to see some recent results, the reader can refer to the following studies: (Adrot and Flaus 2008; Armengol *et al.* 2008; Fagarasan *et al.* 2004; Hamelin and Sauter 2000; Ploix and Adrot 2006; Puig *et al.* 2006; Puig *et al.* 2008; Rambeaux *et al.* 2000; Sainz *et al.* 2002), among others.

This chapter will review the passive approach when considering the nominal model plus the uncertainty on every parameter bounded by intervals. This type of uncertainty modeling provides a type of models known as *interval models*. Noise will also be considered to be unknown but bounded and modeled in a deterministic framework. The use of interval models has received several names depending on the field of application: in circuit analysis, it is known as worst-case analysis (or tolerance analysis); in automatic control, it is known as the *set-membership* approach (also known as the *error-bounded* approach) and in qualitative reasoning, it is termed as semi-quantitative. In the automatic control literature, the *set-membership* approach applied to parameter and state estimation has been discussed extensively in Milanese *et al.* (1996) while their application to control can be found in Bhattacharyya *et al.* (1995) and Ackermann (2002). The worst-case analysis of circuits has been discussed in Kolev (1993) and in several research papers, it appears in circuits journals and conferences.

The remainder of the chapter is organized as follows: section 3.2 introduces some background material on sets and in particular on zonotopes. In section 3.3, the problem set-up considered in this chapter is introduced. Section 3.6 presents the fault detection and isolation procedures using set-based approaches. Section 3.7 presents the application of set-membership methods for a fault detection four-tank case study. Finally, the major conclusions are summarized in section 3.8.

3.2. Notations, definitions and properties

We use I to denote an identity matrix of appropriate dimension. For a matrix X, we denote $\mathrm{tr}(X)$ and $\mathrm{rank}(X)$ as the trace and the rank of X, respectively, X^{-1} and $X^\top$ as the inverse and the transpose of X, respectively, and $X \succ 0$ denotes positive definiteness. We define the following sets $\mathbb{S}^n := \{X \in \mathbb{R}^{n\times n} \mid X = X^T\}$ and $\mathbb{S}^n_{\succ 0} := \{X \in \mathbb{R}^{n\times n} \mid X = X^T, X \succ 0\}$.

DEFINITION 3.1 (Zonotope).– *Given a vector $c \in \mathbb{R}^n$ and a matrix $H \in \mathbb{R}^{n\times r}$, a zonotope $\mathcal{Z} \subset \mathbb{R}^n$ in n-dimensional space is defined as*

$$\mathcal{Z} := \langle c, H\rangle = \{c + Hz, \|z\|_\infty \leq 1\}, \quad [3.1]$$

where c and H are called the center and the segment matrix of the zonotope $\mathcal{Z}$.

With the Minkowski sum $\oplus$, the zonotope in [3.1] can also be defined by $\mathcal{Z} = c \oplus H\mathbf{B}^r$, where $\mathbf{B}^r = [-1, +1]^r \subset \mathbb{R}^r$ is an r-dimensional hypercube. Besides, the following linear properties hold:

$$\langle c_1, H_1\rangle \oplus \langle c_2, H_2\rangle = \langle c_1 + c_2, [H_1 \;\; H_2]\rangle, \quad [3.2a]$$

$$L\langle c, H\rangle = \langle Lc, LH\rangle, \quad [3.2b]$$

where L is an arbitrary matrix of appropriate dimension.

In the following definitions, two measures of the size of a zonotope $\mathcal{Z}$ are introduced from Le *et al.* (2013) and Combastel (2015).

DEFINITION 3.2 (P-radius of a zonotope).– *Given a zonotope $\mathcal{Z} = \langle c, H\rangle \subset \mathbb{R}^n$ and a matrix $P \in \mathbb{S}^n_{\succ 0}$, the P-radius of $\mathcal{Z}$ is defined by*

$$\ell_P(H) := \max_{\|z\|_\infty \leq 1} \|Hz\|_P = \max_{\|z\|_\infty \leq 1} \sqrt{z^\top H^\top P H z}. \quad [3.3]$$

DEFINITION 3.3 (Weighted F-radius of a zonotope).– *Given a zonotope $\mathcal{Z} = \langle c, H\rangle \subset \mathbb{R}^n$ and a matrix $W \in \mathbb{S}^n_{\succ 0}$, the weighted F-radius of $\mathcal{Z}$ based on the weighted Frobenius norm is defined by*

$$\ell_{F,W}(H) := \|H\|_{F,W} = \sqrt{\mathrm{tr}\,(H^\top W H)}. \quad [3.4]$$

For $\mathcal{Z} = \langle c, H\rangle$, the weighted zonotope reduction operator proposed in Combastel (2015) is denoted by $\downarrow_{q,W}(H)$ satisfying the inclusion property $\langle c, H\rangle \subseteq \langle c, \downarrow_{q,W}(H)\rangle$, where $q \geq n$ specifies the maximum number of columns of $\downarrow_{q,W}(H)$ and W is a weighting matrix of appropriate dimension. For notation simplicity, in this chapter, we use $\bar{H}$ to denote this reduction operator.

3.3. Problem statement

3.3.1. *Uncertain discrete-time linear systems*

In this chapter, we consider the class of discrete-time linear time-invariant (LTI) systems subject to disturbances and noise as

$$x(k+1) = Ax(k) + Bu(k) + D_w w(k), \quad [3.5a]$$

$$y(k) = Cx(k) + D_v v(k), \quad [3.5b]$$

where $x \in \mathbb{R}^{n_x}$, $u \in \mathbb{R}^{m_u}$ and $y \in \mathbb{R}^{n_y}$ denote the vectors of system states, known inputs and outputs, respectively. $w \in \mathbb{R}^{m_w}$ and $v \in \mathbb{R}^{m_v}$ denote the vectors of disturbances and measurement noise, respectively. System matrices are given as follows: $A \in \mathbb{R}^{n_x \times n_x}$, $B \in \mathbb{R}^{n_x \times m_u}$, $C \in \mathbb{R}^{n_y \times n_x}$, $D_w \in \mathbb{R}^{n_x \times m_w}$ and $D_v \in \mathbb{R}^{n_y \times n_v}$.

In [3.5], $w(k)$ and $v(k)$ are unknown at each time step $k \in \mathbb{N}$. In order to build a set-based framework, the following assumptions are made.

ASSUMPTION 3.1.– The state disturbance and measurement noise vectors $w(k)$ and $v(k)$ are unknown but bounded by the unitary zonotopes:

$$w(k) \in \langle 0, I_{m_w} \rangle, \; v(k) \in \langle 0, I_{m_v} \rangle, \; \forall k \in \mathbb{N}. \quad [3.6]$$

ASSUMPTION 3.2.– The initial state $x(0)$ is assumed to be bounded by the zonotope $x(0) \in \mathcal{X}(0) := \langle c(0), H(0) \rangle$.

3.3.2. *Set-based methods*

Taking into account uncertainties in the considered LTI model, set-based methods are usually used for explicitly describing the uncertainty propagation along system dynamics. In this chapter, we consider two set-based methods, both considering the propagation of uncertainties using zonotopes.

1) Set-membership state estimation has been discussed over the last decade (see e.g. (Puig *et al.* 2001; Alamo *et al.* 2005, 2008; Wang *et al.* 2018)). The idea behind the set-membership approach is to use the customized geometrical set to bound uncertain states during iterations. The alternative geometrical sets include intervals,

ellipsoids, polytopes, zonotopes and so on. Due to their low computational complexity, zonotopes, a type of symmetrical set, are usually chosen, for instance in Alamo *et al.* (2005).

Algorithm 3.1. Set-membership state estimator

1: $\mathcal{X}_k^e \Leftarrow \mathcal{X}_0$
2: **for** $k = 1$ to N **do**
3: Compute the set of predicted states $\mathcal{X}^p(k)$ using [3.5a] and uncertainty bounds
4: Compute the set states consistent with the outputs $\mathcal{X}^y(k)$ using [3.5b] and uncertainty bounds
5: Compute the set of estimated states $\mathcal{X}(k)^e = \mathcal{X}(k)^p \cap \mathcal{X}^y(k)$
6: $k \leftarrow k + 1$
7: **end for**

2) Set-based observer is another alternative set-based method (see e.g. Raïssi *et al.* (2012); Combastel (2015). The idea is to combine the traditional observer structure and set theory. By using the standard Luenberger observer structure, the zonotopic set enclosed uncertain states are propagated with unknown-but-bounded disturbances and measurement noise.

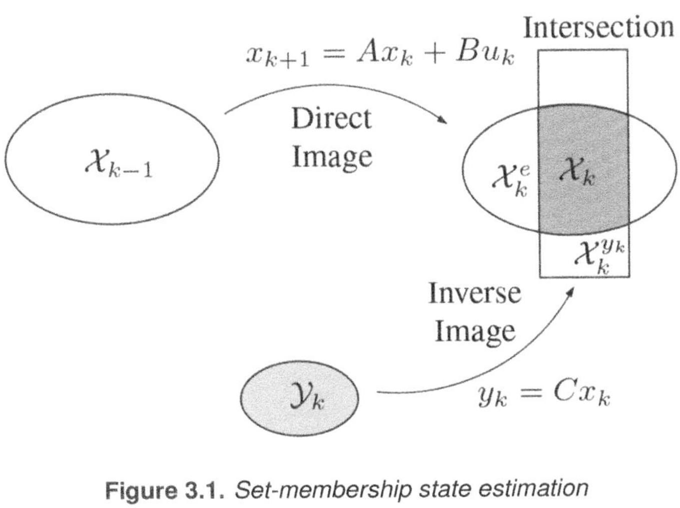

Figure 3.1. *Set-membership state estimation*

Algorithm 3.2. Set-based observer

$k \leftarrow 1$

$\mathcal{X}_k^e \Leftarrow \mathcal{X}_0$

while $N \leq k$ **do**

  Obtain and store input–output data $\{u(k), y(k)\}$

  Compute the set estimated states $\mathcal{X}^e(k)$ using [3.5a] and uncertainty bounds

  Compute the set of estimated outputs $\mathcal{Y}^e(k)$ using [3.5b] and uncertainty bounds

  $k \leftarrow k + 1$

end while

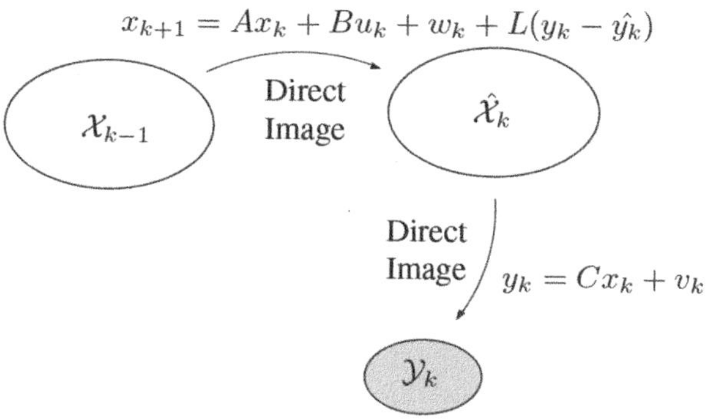

Figure 3.2. *Interval observer*

3.3.3. *FDI problem statement*

Let us consider the system model [3.5] again, but now including faults:

$$x(k+1) = Ax(k) + Bu(k) + D_w w(k) + F_u f_u(k), \quad [3.7a]$$

$$y(k) = Cx(k) + D_v v(k) + F_y f_y(k), \quad [3.7b]$$

where $f \in \mathbb{R}^{m_f}$ and $F \in \mathbb{R}^{n_y \times m_f}$ denote the vector and the magnitude matrix of actuator (subindex u) and sensor (subindex y) faults, respectively.

In this chapter, we will introduce how to use set-based estimation methods described faults to detect and isolate faults considering uncertainty in a robust manner.

3.4. Proposed techniques

In this section, we introduce the structures of two set-based approaches and their relationship. We then present several design methods for these observers.

In the following, two set-based estimation approaches are formulated. One is for the set-membership approach introduced from Alamo *et al.* (2005) and another one is for the zonotopic observer introduced from Combastel (2015).

3.4.1. *Set-membership approach*

We first introduce the set-membership approach for a nominal discrete-time system [3.5]. In this approach, the uncertain state at each time step k can be bounded by an explicit zonotopic set.

PROPOSITION 3.1 (State-bounding zonotope).– Given the system [3.5], suppose that assumption 3.1 holds, and that $x(k-1) \in \langle c(k-1), H(k-1)\rangle, \forall k \in \mathbb{N}$. The following inclusion then holds for every correction matrix $\Lambda \in \mathbb{R}^{n_x \times n_y}$:

$$x(k) \in \langle c(k), H(k)\rangle, \quad [3.8]$$

where

$$c(k) = (I - \Lambda C)\Big(Ac(k-1) + Bu(k-1)\Big) + \Lambda y(k), \quad [3.9a]$$

$$H(k) = \big[(I - \Lambda C)AH(k-1),\, (I - \Lambda C)D_w,\, \Lambda D_v\big]. \quad [3.9b]$$

PROOF.– Given $x(k-1) \in \langle c(k-1), H_i(k-1)\rangle$, by definition 3.1, there exists a vector $\theta(k-1)$ with $\|\theta(k-1)\|_\infty \leq 1$ such that

$$x(k-1) = c(k-1) + H(k-1)\theta(k-1).$$

According to the system dynamics in [3.5], in the prediction step, we can propagate

$$\begin{aligned} x(k) &= Ax(k-1) + Bu(k-1) + w(k-1) \\ &= A\left(c(k-1) + H(k-1)\theta(k-1)\right) + Bu(k-1) + w_(k-1). \end{aligned} \quad [3.10]$$

From assumption 3.1, there exists a vector $\varpi(k-1)$ with $\|\varpi(k-1)\|_\infty \leq 1$ such that $w(k-1) = D_w\varpi(k-1)$. Thus, from [3.10], we derive

$$x(k) = A\left(c(k-1) + H(k-1)\theta(k-1)\right) Bu(k-1) + D_w\varpi(k-1).$$

Let us set $\hat{c}(k) = Ac(k-1)+Bu(k-1)$, $R(k) = [AH(k-1), D_w]$ and $\eta(k-1) = \left[\theta(k-1)^\top, \varpi(k-1)^\top\right]^\top$, where $\|\eta(k-1)\|_\infty \leq 1$. Then, we have

$$x(k) = \hat{c}(k) + R(k)\eta(k-1). \qquad [3.11]$$

From assumption 3.1, there exists a vector $\sigma(k)$ with $\|\sigma(k)\|_\infty \leq 1$ such that $v(k) = D_v\sigma(k)$. From the output equation of [3.5], we have

$$y(k) - Cx(k) - D_v\sigma(k) = 0. \qquad [3.12]$$

Thus, by replacing $x(k)$ in [3.12] with the expression in [3.11], we obtain

$$y(k) - C\hat{c}(k) - CR(k)\eta(k-1) - D_v\sigma(k) = 0.$$

Pre-multiplying by Λ_i and rearranging the terms of the above equation yields

$$\Lambda y(k) - \Lambda C\hat{c}(k) - \Lambda CR(k)\eta(k-1) - \Lambda D_v\sigma(k) = 0. \qquad [3.13]$$

Finally, in the correction step, we add [3.13] to the right side of [3.11], obtaining

$$\begin{aligned} x(k) &= \hat{c}(k) + R(k)\eta(k-1) + \Lambda y(k) - \Lambda C\hat{c}(k) \\ &\quad - \Lambda CR(k)\eta(k-1) - \Lambda D_v\sigma(k). \end{aligned}$$

By setting $c(k)$ and $H(k)$ as in [3.9], the above equation becomes

$$x(k) = c(k) + H(k)\begin{bmatrix}\eta(k-1)\\ -\sigma(k)\end{bmatrix} = c(k) + H(k)\begin{bmatrix}\eta(k-1)\\ \sigma(k)\end{bmatrix}.$$

Since $\|\eta(k-1)\|_\infty \leq 1$ and $\|\sigma_i(k)\|_\infty \leq 1$, we can conclude that $x(k) \in \langle c(k), H(k)\rangle$ as in [3.8]. □

3.4.2. *Zonotopic observer*

We next introduce a zonotopic observer, which is built based on a prediction-type observer structure. For the system [3.5], the introduction of an observer gain $G(k) \in \mathbb{R}^{n_x \times n_y}$ gives the prediction-type observer structure as

$$x(k+1) = Ax(k) + Bu(k) + D_w w(k) + G(k)(y(k) - Cx(k) - D_v v(k)). \qquad [3.14]$$

Based on [3.14], we can define the prediction-type zonotopic observer below.

PROPOSITION 3.2 (Prediction-type zonotopic observer).– Given the system [3.5], suppose that assumptions 3.1 and 3.2 hold. Let us recursively define the center and the segment matrix for the prediction-type zonotopic observer by

$$c(k+1) = (A - G(k)C)\, c(k) + Bu(k) + G(k)y(k), \qquad [3.15a]$$

$$H(k+1) = \left[(A - G(k)C)\, \bar{H}(k),\, D_w,\, G(k)D_v\right]. \qquad [3.15b]$$

Then, the state inclusion holds

$$x(k) \in \mathcal{X}(k) = \langle c(k), H(k)\rangle, \forall k \in \mathbb{N}. \qquad [3.16]$$

PROOF.– Consider that $x(k) \in \langle c(k), H(k)\rangle$ at a time k. As introduced before, with the reduction operator, we also have $x(k) \in \langle c(k), \bar{H}(k)\rangle$. Since $w(k) \in \langle 0, I_{m_w}\rangle$, $v(k) \in \langle 0, I_{m_v}\rangle$, from [3.14], it follows

$$\begin{aligned} x(k+1) &\in \langle c(k+1), H(k+1)\rangle \\ &= ((A - G(k)C)\langle c(k), \bar{H}(k)\rangle) \oplus (B\langle u(k), 0\rangle) \oplus (G(k)\langle y(k), 0\rangle) \\ &\quad \oplus (D_w\langle 0, I_{m_w}\rangle) \oplus ((-G(k)D_v)\langle 0, I_{m_v}\rangle). \end{aligned}$$

By using properties given in [3.2] in the above equation, we obtain $c(k+1)$ and $H(k+1)$ as in [3.15]. □

3.4.3. *Relationship between set-based methods*

As discussed in Wang *et al.* (2018), the relationship between the set-membership approach and the zonotopic observer has been built for a more general system called descriptor systems. The relationship also applies to the dynamic system in the form of [3.5]. On the basis of this, we now discuss the relationship between these two set-based methods.

Comparing the state bounding zonotopes defined in propositions 3.1 and 3.2, the measurement output $y(k)$ is used in both zonotopes to correct the predicted state-bounding zonotopes. However, the output $y(k)$ is used to find $x(k) \in \langle c(k), H(k)\rangle$ in the set-membership approach, while $x(k+1) \in \langle c(k+1), H(k+1)\rangle$ in the zonotopic observer.

To find the relationship between two set-based methods, we need to consider a current estimation-type observer structure for the system [3.5] as follows:

$$x(k+1) = Ax(k) + Bu(k) + D_w w(k) \qquad [3.17]$$

$$+ \tilde{G}(k)\left(y(k+1) - C\hat{x}(k+1) - D_v v(k+1)\right), \qquad [3.18]$$

where $\tilde{G}(k) \in \mathbb{R}^{n_x \times n_y}$ is a time-varying observer gain for the current estimation-type zonotopic observer. $\hat{x}(k+1)$ denotes a predicted state from the previous observed state $x(t)$ that can be computed by

$$\hat{x}(k+1) = Ax(k) + Bu(k) + D_w w(k). \qquad [3.19]$$

PROPOSITION 3.3.– Given the system [3.5], suppose that assumptions 3.1 and 3.2 hold. The set-membership approach introduced in proposition 3.1 is equivalent to the current estimation-type zonotopic observer in the structure of [3.17].

PROOF.– First, together with the current estimation-type observer in [3.17]–[3.19], we can obtain

$$\begin{aligned} x(k+1) &= (I - \tilde{G}(k)C)Ax(k) + (I - \tilde{G}(k)C)Bu(k) \\ &\quad + (I - \tilde{G}(k)C)D_w w(k) + \tilde{G}(k)y(k+1) - \tilde{G}(k)D_v v(k+1). \end{aligned}$$

Since $x(k) \in \langle c(k), \bar{H}(k) \rangle$, $w(k) \in \langle 0, I_{m_w} \rangle$ and $v(k) \in \langle 0, I_{m_v} \rangle$, from the above equation, we get

$$\begin{aligned} x(k+1) &\in \langle c(k+1), H(k+1) \rangle \\ &= \Big((I - \tilde{G}(k)C)A\langle c(k), \bar{H}(k) \rangle \Big) \oplus \Big((I - \tilde{G}(k)C)B\langle u(k), 0 \rangle \Big) \\ &\quad \oplus \Big((I - \tilde{G}(k)C)D_w\langle 0, I_{m_w} \rangle \Big) \oplus \Big(\tilde{G}(k)\langle y(k+1), 0 \rangle \Big) \\ &\quad \oplus \Big((-\tilde{G}(k)D_v)\langle 0, I_{m_v} \rangle \Big). \end{aligned}$$

Also using properties in [3.2], the above equation gives a current estimation-type zonotopic observer. It is obvious that this observer is equivalent to the set-membership approach defined in proposition 3.1 at time step $k+1$. □

3.5. Design methods

For set-based methods, the propagated uncertainties determine the size of the state-bounding zonotopes. First, we can establish the robustness conditions to minimize the size of those zonotopes. In the literature, we introduce two robustness conditions for the set-membership approach and the zonotopic observer, respectively.

3.5.1. *Robustness conditions*

PROPOSITION 3.4 (Robust Correction Matrix).– Given the state-bounding zonotope $\mathcal{X}(k) = \langle c(k), H(k) \rangle$, $\forall k \in \mathbb{N}$ as in proposition 3.1, two scalars $0 < \gamma_c < 1$ and

$\epsilon > 0$. If there exist matrices $P \in \mathbb{S}^{n_x}_{\succ 0}$, $Y \in \mathbb{R}^{n_x \times n_y}$, diagonal matrices $\Gamma \in \mathbb{S}^{m_w}_{\succeq 0}$ and $\Upsilon \in \mathbb{S}^{m_v}_{\succeq 0}$ such that

$$\mathrm{tr}(\Gamma) + \mathrm{tr}(\Upsilon) < \epsilon, \qquad [3.20a]$$

$$\begin{bmatrix} \gamma_c P & \star & \star & \star \\ PA - YCA & P & \star & \star \\ 0 & D_w^\top P - D_w^\top C^\top Y^\top & \Gamma & \star \\ 0 & D_v^\top Y^\top & 0 & \Upsilon \end{bmatrix} \succ 0, \qquad [3.20b]$$

then it is guaranteed that

$$V(\mathcal{X}(k)) \le \gamma_c V(\mathcal{X}(k-1)) + \epsilon, \; \forall k \in \mathbb{N}_+, \qquad [3.21]$$

leading to $V(\mathcal{X}(\infty)) \le \frac{\epsilon}{1-\gamma_c}$ when $k \to \infty$. Besides, the correction matrix can be obtained by $\lambda = P^{-1}Y$.

PROOF.– The size of the zonotope $\mathcal{X}(k)$ can be measured by the P-radius. With $P \in \mathbb{S}^{n_x}_{\succ 0}$, for every $H(k-1)$ and $0 < \gamma_c < 1$, the condition [3.21] is equivalent to

$$\max_{\|\phi\|_\infty \le 1} \|H(k)\phi\|_P^2 - \max_{\|\theta\|_\infty \le 1} \gamma_c \|H(k-1)\theta\|_P^2 - \epsilon \le 0. \qquad [3.22]$$

Since $\max_\theta \|H(k-1)\theta\|_P^2 \ge \|H(k-1)\theta\|_P^2$ for any $\|\theta\|_\infty \le 1$, we obtain the following sufficient condition of [3.22]

$$\max_{\|\phi\|_\infty \le 1} \|H(k)\phi\|_P^2 - \gamma_c \|H(k-1)\theta\|_P^2 - \epsilon \le 0. \qquad [3.23]$$

Denote $\phi = \left[\theta^\top, \varpi^\top, \sigma^\top\right]^\top$ and $Y = R\Lambda$, then

$$\left[(P - YC)A, \, (P - YC)D_w, \, YD_v\right] = R. \qquad [3.24]$$

With this notation, we have

$$H(k)\phi = P^{-1}R\left[(H(k-1)\theta)^\top, \varpi^\top, \sigma^\top\right]^\top.$$

Therefore, we rewrite [3.23] as

$$\begin{bmatrix} H(k-1)\theta \\ \varpi \\ \sigma \end{bmatrix}^\top R^\top P^{-1} R \begin{bmatrix} H(k-1)\theta \\ \varpi \\ \sigma \end{bmatrix} - \gamma_c (H(k-1)\theta)^\top P H(k-1)\theta - \epsilon < 0, \qquad [3.25]$$

for any $\|\phi\|_\infty \leq 1$. If $\Gamma \in \mathbb{S}^{m_w}_{\succeq 0}$ and $\Upsilon \in \mathbb{S}^{m_v}_{\succeq 0}$ are diagonal matrices, then we have

$$\mathrm{tr}(\Gamma) - \varpi^\top \Gamma \varpi \geq 0, \ \forall \|\varpi\|_\infty \leq 1, \quad [3.26a]$$

$$\mathrm{tr}(\Upsilon) - \sigma^\top \Upsilon \sigma \geq 0, \ \forall \|\sigma\|_\infty \leq 1. \quad [3.26b]$$

By adding [3.26] to [3.25], we can obtain a sufficient condition

$$\begin{bmatrix} H(k-1)\theta \\ \varpi \\ \sigma \end{bmatrix}^\top R^\top P^{-1} R \begin{bmatrix} H(k-1)\theta \\ \varpi \\ \sigma \end{bmatrix} - \gamma_c (H(k-1)\theta)^\top P H(k-1)\theta - \epsilon$$
$$+ \mathrm{tr}(\Gamma) - \varpi^\top \Gamma \varpi + \mathrm{tr}(\Upsilon) - \sigma^\top \Upsilon \sigma < 0.$$

If [3.20a] is satisfied, then we obtain

$$\begin{bmatrix} H(k-1)\theta \\ \varpi \\ \sigma \end{bmatrix}^\top \left(R^\top P^{-1} R - \begin{bmatrix} \gamma_c P & 0 & 0 \\ 0 & \Gamma & 0 \\ 0 & 0 & \Upsilon \end{bmatrix} \right) \begin{bmatrix} H(k-1)\theta \\ \varpi \\ \sigma \end{bmatrix} < 0.$$

From the above inequality, we have a sufficient condition

$$\begin{bmatrix} \gamma_c P & 0 & 0 \\ 0 & \Gamma & 0 \\ 0 & 0 & \Upsilon \end{bmatrix} - R^\top P^{-1} R \succ 0,$$

and by applying the Schur complement, we obtain

$$\begin{bmatrix} \gamma_c P & \star & \star & \star \\ 0 & \Gamma & \star & \star \\ 0 & 0 & \Upsilon & \star \\ (P - YC)A & (P - YC)D_w & YD_v & P \end{bmatrix} \succ 0.$$

Finally, we obtain [3.20b] through a linear coordinate transformation by the matrix $T = \begin{bmatrix} I & 0 & 0 & 0 \\ 0 & 0 & I & 0 \\ 0 & 0 & 0 & I \\ 0 & I & 0 & 0 \end{bmatrix}$ applied to the above inequality. □

PROPOSITION 3.5 (Zonotopic Kalman Gain (Combastel 2015)).– Given the zonotopic observer in proposition 3.2 and a weighting matrix $W \in \mathbb{S}^{n_r}_{\succ 0}$, the zonotopic Kalman gain $G^*(k) = \arg\min_G \|H(k+1)\|^2_{F,W}$ is given as follows:

$$G^*(k) = AK^*(k), \qquad [3.27a]$$

$$K^*(k) = L(k)S(k)^{-1}, \qquad [3.27b]$$

$$L(k) = \bar{H}(k)\bar{H}(k)^\top C^\top, \qquad [3.27c]$$

$$S(k) = C\bar{H}(k)\bar{H}(k)^\top C^\top + D_v D_v^\top. \qquad [3.27d]$$

PROOF.– The criterion $J = \ell_{F,W}(H(k+1))^2$ is convex with respect to $G(k)$. $G^*(k)$ is a value of $G(k)$ such that $\frac{\partial J}{\partial G} = 0$. □

REMARK 3.1.– From the proof of proposition 3.5, it is clear that the zonotopic Kalman gain does not depend on the weighting matrix W. Therefore, W can be set as an arbitrary matrix (i.e. an identity matrix).

The conditions in proposition 3.4 can also be applied to build the robustness condition for the zonotopic observer.

COROLLARY 3.1 (Robust Zonotopic Observer Gain).– Given the zonotopic observer $\mathcal{X}(k) = \langle c(k), H(k) \rangle$, $\forall k \in \mathbb{N}$ as in proposition 3.2, two scalars $0 < \gamma_o < 1$ and $\epsilon_o > 0$. If there exist matrices $P \in \mathbb{S}^{n_x}_{\succ 0}$ and $Y \in \mathbb{R}^{n_x \times n_y}$ such that

$$\begin{bmatrix} \gamma_o P & \star & \star & \star \\ 0 & D_w^\top D_w & \star & \star \\ 0 & 0 & D_v^\top D_v & \star \\ PA - YC & PD_w & YD_v & P \end{bmatrix} \succ 0, \qquad [3.28]$$

then it is guaranteed that

$$V(\mathcal{X}(k+1)) \leq \gamma_o V(\mathcal{X}(k)) + \epsilon_o, \; \forall k \in \mathbb{N}, \qquad [3.29]$$

leading to $V(\mathcal{X}(\infty)) \leq \frac{\epsilon_o}{1-\gamma_o}$ when $k \to \infty$. Besides, the zonotopic observer gain can be obtained by $G = P^{-1}Y$.

PROOF.– The size of the zonotope $\mathcal{X}(k)$ in [3.15] can also be measured by the P-radius. Similar to the proof of the proposition 3.4, we consider the condition [3.29], where ϵ_o is the slack term affected by uncertainties that can be assigned as

$$\epsilon_o = \max_{\|\varpi\|_\infty \leq 1} \|D_w \varpi\|_2^2 + \max_{\|\sigma\|_\infty \leq 1} \|D_v \sigma\|_2^2. \qquad [3.30]$$

Then, [3.29] can be rewritten as

$$\max_{\|\phi\|_\infty \leq 1} \|H(k+1)\phi\|_P^2 - \max_{\|\theta\|_\infty \leq 1} \gamma_o \|H(k)\theta\|_P^2 - \max_{\|\varpi\|_\infty \leq 1} \|D_w \varpi\|_2^2 - \max_{\|\sigma\|_\infty \leq 1} \|D_v \sigma\|_2^2 \leq 0.$$

For any unitary vectors $\phi = [\theta^\top, \varpi^\top, \sigma^\top]^\top, \theta, \varpi, \sigma$, a sufficient condition of previous inequality can be found as follows:

$$\begin{bmatrix} \theta \\ \varpi \\ \sigma \end{bmatrix}^\top H(k+1)^\top P H(k+1) \begin{bmatrix} \theta \\ \varpi \\ \sigma \end{bmatrix} - \theta^\top H(k)^\top \gamma_o P H(k) \theta$$
$$- \varpi^\top D_w^\top D_w \varpi - \sigma^\top D_v^\top D_v \sigma < 0.$$

For set $\bar{\theta} = H(k)\theta$, it follows

$$\begin{bmatrix} \bar{\theta} \\ \varpi \\ \sigma \end{bmatrix}^\top \Psi \begin{bmatrix} \bar{\theta} \\ \varpi \\ \sigma \end{bmatrix} - \begin{bmatrix} \bar{\theta} \\ \varpi \\ \sigma \end{bmatrix}^\top \begin{bmatrix} \gamma_o P & 0 & 0 \\ 0 & D_w^\top D_w & 0 \\ 0 & 0 & D_v^\top D_v \end{bmatrix} \begin{bmatrix} \bar{\theta} \\ \varpi \\ \sigma \end{bmatrix} < 0,$$

with $\Psi = \begin{bmatrix} A - GC & D_w & GD_v \end{bmatrix}^\top P \begin{bmatrix} A - GC & D_w & GD_v \end{bmatrix}$. Based on the definition of the negative definite matrix, we have

$$\begin{bmatrix} \gamma_o P & 0 & 0 \\ 0 & D_w^\top D_w & 0 \\ 0 & 0 & D_v^\top D_v \end{bmatrix}$$
$$- \begin{bmatrix} P(A - GC) & PD_w & PGD_v \end{bmatrix}^\top P^{-1} \begin{bmatrix} P(A - GC) & PD_w & PGD_v \end{bmatrix} \prec 0.$$

By using the Schur complement, we can obtain the LMI condition [3.28] by setting $Y = PG$. □

3.5.2. *Fault sensitivity condition*

For fault detection, the optimal correction matrix and two zonotopic observer gains can be applied. Furthermore, if we have more knowledge/information of occurred faults, a second condition of fault sensitivity can also be introduced in the design of a fault detection gain.

To implement an FD scheme based on zonotopic observer, a residual zonotope can be defined as follows.

COROLLARY 3.2.– Given uncertain state enclosed in the zonotope $x(k) \in \langle c(k), H(k) \rangle, \forall k \in \mathbb{N}$ in the form of [3.16]. Then, the corresponding residual zonotope $\mathcal{R}(k) = \langle c_r(k), H_r(k) \rangle, \forall k \in \mathbb{N}$ can be formulated as

$$c_r(k) = M\left(y(k) - Cc(k)\right), \qquad [3.31a]$$

$$H_r(k) = \begin{bmatrix} MCH(k) & MD_v \end{bmatrix}, \qquad [3.31b]$$

where $M \in \mathbb{R}^{n_y \times n_y}$ denotes the gain of the residual zonotope.

PROOF.– The residual zonotope is directly obtained from a linear projection of the state bounding zonotope $x(k) \in \langle c(k), H(k) \rangle$. The details are omitted here. □

For a faulty system in [3.7], a sensor fault sensitivity condition is given as follows.

PROPOSITION 3.6 (Sensor Fault Sensitivity Condition).– Consider the zonotopic observer in proposition 3.2. If there exist matrices $P \in \mathbb{S}_{\succ 0}^{n_x}$, $Y \in \mathbb{R}^{n_x \times n_y}$, $N \in \mathbb{R}^{n_y \times n_y}$, and scalars $\alpha_s > 0$, $\beta_s > 0$ such that

$$\begin{bmatrix} \alpha_s P & \star & \star \\ \alpha_s A^\top P - \alpha_s C^\top Y^\top & \alpha_s P + C^\top N C & \star \\ -\alpha_s F^\top Y^\top & F^\top N^\top C & F^\top N F - \beta_s^2 I \end{bmatrix} \succ 0, \qquad [3.32]$$

then the $\mathcal{H}_-$ fault sensitivity condition holds:

$$\sum_{k=0}^{\infty} c_{r_f}(k)^\top c_{r_f}(k) \geq \beta_s^2 \sum_{k=0}^{\infty} f(k)^\top f(k), \qquad [3.33]$$

where c_{r_f} denotes the center of residual zonotope [3.31] affected by sensor faults. Besides, the gains can be obtained by $G = P^{-1}Y$ and $M = N^{\frac{1}{2}}$.

PROOF.– Considering that there are no disturbances and noise, $w(k) = 0$ and $v(k) = 0$ for $k \in \mathbb{N}$. The state estimation error $e_f(k)$, only affected by the sensor faults at time step k, can be defined to be $e_f(k) = x(k) - c(k)$. Therefore, it can be also verified that the error dynamics can be formulated as

$$e_f(k+1) = (A - LC)e_f(k) - GFf(k). \qquad [3.34]$$

Due to this there are no disturbances and noise, and the residual zonotope $\mathcal{R}(k)$ is aggregated into its center $c_{r_f}(k)$ at time step k. Therefore, the center of the residual zonotope $c_{r_f}(k)$, only affected by the sensor fault vector $f(k)$, can be formulated as

$$\begin{aligned} c_{r_f}(k) &= M(Cx(k) + Ff(k) - Cc(k)) \\ &= MCe_f(k) + MFf(k). \end{aligned} \qquad [3.35]$$

From the condition [3.33], we can define a criterion function as

$$J_s = \sum_{k=0}^{T-1} \left(c_{r_f}(k)^\top c_{r_f}(k) - \beta_s^2 f(k)^\top f(k) \right), \qquad [3.36]$$

where T is an arbitrary time horizon.

Furthermore, there exists a matrix $Q_s \in \mathbb{S}^{n_x}_{\succ 0}$ such that a Lyapunov candidate function can be chosen as

$$V_f(k) = e_f(k)^\top Q_s e_f(k), \tag{3.37}$$

such that

$$\begin{aligned}\Delta V_f(k) &= V_f(k+1) - V_f(k)\\ &= e_f(k)^\top \left((A-GC)^\top Q_s(A-GC) - Q_s\right) e_f(k)\\ &\quad - 2e_f(k)^\top (A-GC)^\top Q_s GFf(k)\\ &\quad + f(k)^\top (GF)^\top Q_s(GF)f(k).\end{aligned}$$

For any non-zero $f(k) \in \mathcal{L}_2[0,\infty)$ and zero initial condition $e_f(0) = 0$, [3.36] is equivalent to

$$J_s = \sum_{k=0}^{T-1} \left(c_{r_f}(k)^\top c_{r_f}(k) - \beta_s^2 f(k)^\top f(k) - \Delta V_f(k)\right) + V_f(T).$$

The condition [3.33] can then be satisfied whenever $J_s \geq 0$. Therefore, a sufficient condition is given as

$$c_{r_f}(k)^\top c_{r_f}(k) - \beta_s^2 f(k)^\top f(k) - \Delta V_f(k) \succ 0, \tag{3.38}$$

By substituting $c_{r_f}(k)$ in [3.38] by [3.35], [3.38] becomes

$$\begin{bmatrix} e_f(k) \\ f(k) \end{bmatrix}^\top \begin{bmatrix} \Phi_1^s & \star \\ \Phi_2^s & \Phi_3^s \end{bmatrix} \begin{bmatrix} e_f(k) \\ f(k) \end{bmatrix} \succ 0, \tag{3.39}$$

with

$$\begin{aligned}\Phi_1^s &= C^\top M^\top MC - (A-GC)^\top Q_s(A-GC) + Q_s,\\ \Phi_2^s &= (GF)^\top Q_s(A-GC) + F^\top MM^\top C,\\ \Phi_3^s &= F^\top M^\top MF - \beta_s^2 I - (GF)^\top Q_s(GF).\end{aligned}$$

By using the Schur complement, from [3.39] we can obtain

$$\begin{bmatrix} Q_s & \star & \star \\ A^\top Q_s - C^\top G^\top Q_s & Q_s + C^\top M^\top MC & \star \\ -F^\top G^\top Q_s & F^\top MM^\top C & F^\top M^\top MF - \beta_s^2 I \end{bmatrix} \succ 0. \tag{3.40}$$

Let us set $N = M^\top M$. Without loss of generality, there exists a positive scalar α_s such that the matrix Q_s is linked with the positive matrix P in the P-radius of the zonotope by $Q_s = \alpha_s P$. Then, [3.40] is equivalent to [3.32] with $Y = PG$. □

3.6. Fault detection and isolation procedures

We now present how the previous set-based approaches can be used for robust fault detection and isolation.

3.6.1. *Fault detection*

3.6.1.1. *Set-membership estimation*

The use of set-membership state estimation in fault detection is very straightforward. The existence of a fault will be detected through the set-membership state estimation algorithm based on set-computations (algorithm 3.3) in the intersection step. Then, assuming that the actual system satisfies [3.5] under non-faulty operating conditions, the set-membership algorithm based on set-computations is correctly initialized with the initial condition given by $\mathcal{X}_0$; for a given sequence of measured inputs $(\mathbf{u}_j)_0^{k-1}$ and outputs $(\mathbf{y}_j)_0^k$ of the actual system, a fault is said to have occurred if

$$\mathcal{X}_k = \mathcal{X}_k^e \cap \mathcal{X}_k^{y_k} = \emptyset, \quad [3.41]$$

at any time instant k.

Algorithm 3.3. Fault detection using set-membership state estimator

1: $\mathcal{X}_k^e \Leftarrow \mathcal{X}_0$
2: **for** $k = 1$ to N **do**
3: Compute the set of predicted states, $\mathcal{X}_k^p$
4: Compute the set of consistent states, $\mathcal{X}_k^c$
5: Compute the set of uncertain states, $\mathcal{X}_k^e = \mathcal{X}_k^p \cap \mathcal{X}_k^c$
6: **if** $\mathcal{X}_k^e = \emptyset$ **then**
7: Exit (Fault detected)
8: **end if**
9: **end for**

When a fault occurs, a recovery strategy is needed. One possibility consists of resetting the set $\mathcal{X}_k^e$ to a size which is guaranteed to capture the true state of the system after the fault has been detected.

3.6.1.2. *Zonotopic observer*

In the case of assuming bounded noise, the measurement can be considered to be in the interval $[y_k]$. The fault detection test can then be stated as

$$[y_k] \cap \mathcal{Y}_k^e \neq \emptyset, \qquad [3.42]$$

where $\mathcal{Y}_k^e$ is set of predicted outputs that can be obtained using algorithm 3.3.

Algorithm 3.4 implements fault detection using interval observers and the fault detection test presented in [3.42].

Algorithm 3.4. Fault detection using interval observers

$fault \leftarrow FALSE$
$k \leftarrow 0$
$\mathcal{X}_k^e \Leftarrow \mathcal{X}_0$
while $fault = FALSE$ **do**
    Obtain input–output data $\{u(k), y(k)\}$
    Compute the set of estimated outputs, $\mathcal{Y}_k^e$ using algorithm 3.2
    if $\mathcal{Y}_k \cap \mathcal{Y}_k^e = \emptyset$ **then**
        $fault \leftarrow TRUE$
    end if
    $k \leftarrow k + 1$
end while

3.6.2. *Fault isolation*

Using the previous fault detection schemes, fault isolation can be implemented by means of a bank of UIO, as the standard procedure introduced in Blanke *et al.* (2016). For multiple faults, each observer is designed to be sensitive to all faults except one.

The decoupling of actuator faults in the state estimation schemes presented above can be done in several ways. Here, we will use an UIO scheme that considers an additional matrix into the state estimation equation, which is then used for decoupling the fault (unknown input) effect on the state estimation error, according to Chen and Patton (1999), or adequately transforming the system [3.44] into a system without the fault (unknown input), as in Keller and Darouach (1999). Anyway, in all the UIO methods, the necessary condition for the existence of a solution to the unknown input decoupling problem is given as in Chen and Patton (1999):

$$rank(CF_u) = rank(F_u) = n_{f_u}, \qquad [3.43]$$

If the condition [3.43] is satisfied, then it is possible to calculate $O = (CF_u)^+ = [(CF_u)^T CF_u]^{-1}(CF_u)^T$, where $(\cdot)^+$ denotes the pseudoinverse of its argument. Then, by multiplying [3.7a] by O, it is straightforward to show that

$$f_u(k) = O[y(k+1) - C(Ax(k) + Bu(k) + D_w w(k)) - D_v v(k+1)]. \qquad [3.44]$$

Then, substituting [3.44] into [3.7] leads to

$$x(k+1) = \bar{A}x(k) + \bar{B}u(k) + \bar{F}_u y(k+1) + D_w \bar{w}(k), \qquad [3.45]$$

where:

$$\begin{aligned} \bar{A} &= (I - F_u OC)A, \\ \bar{B} &= (I - F_u OC)B, \\ \bar{F}_u &= F_u O D_v, \\ \bar{w}(k) &= (I - F_u OC)D_w w(k) - \bar{F}_u v(k+1), \end{aligned}$$

that is decoupled with respect to f_u.

3.7. Application example: quadruple-tank system

In this section, we provide an example with the quadruple-tank system shown in Figure 3.3. As presented in Pourasghar *et al.* (2019), a discrete-time system model of this system is formulated in the form of [3.5] with system matrices:

$$A = \begin{bmatrix} 0.9841 & 0 & 0.0419 & 0 \\ 0 & 0.9889 & 0 & 0.0333 \\ 0 & 0 & 0.9581 & 0 \\ 0 & 0 & 0 & 0.9667 \end{bmatrix}, B = \begin{bmatrix} 0.2102 & 0 \\ 0 & 0.0628 \\ 0 & 0.0479 \\ 0.0094 & 0 \end{bmatrix},$$

$$C = \begin{bmatrix} 0.5 & 0 & 0 & 0 \\ 0 & 0.5 & 0 & 0 \end{bmatrix}, D_w = \begin{bmatrix} 0.01 & 0 & 0 & 0 \\ 0 & 0.01 & 0 & 0 \\ 0 & 0 & 0.01 & 0 \\ 0 & 0 & 0 & 0.01 \end{bmatrix}, D_v = \begin{bmatrix} 0.1 & 0 \\ 0 & 0.1 \end{bmatrix},$$

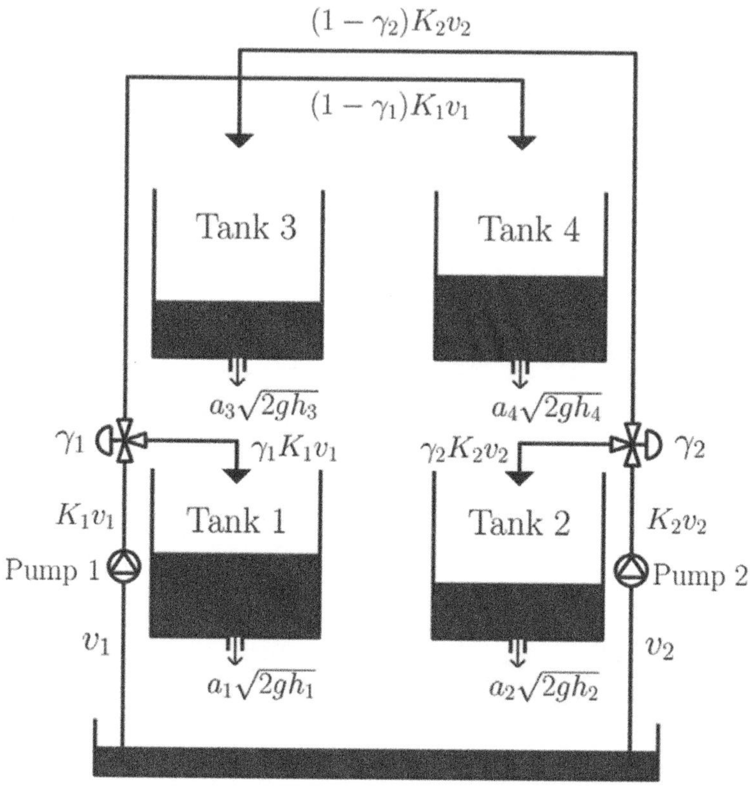

Figure 3.3. *The quadruple-tank system*

For this quadruple-tank system, two normalized sensor faults $f = [f_1, f_2]^{\top}$ are considered, where $F_s = 10I$ and $f_1, f_2 \in [-1, +1]$. These sensor faults are assumed to occur in the system during the time period $200 \leq k \leq 600$. With this consideration, the simulation of quadruple-tank system has been carried out for 1000 time steps. As a result, the closed-loop data of system states and measurement outputs have been collected as shown in Figure 3.4.

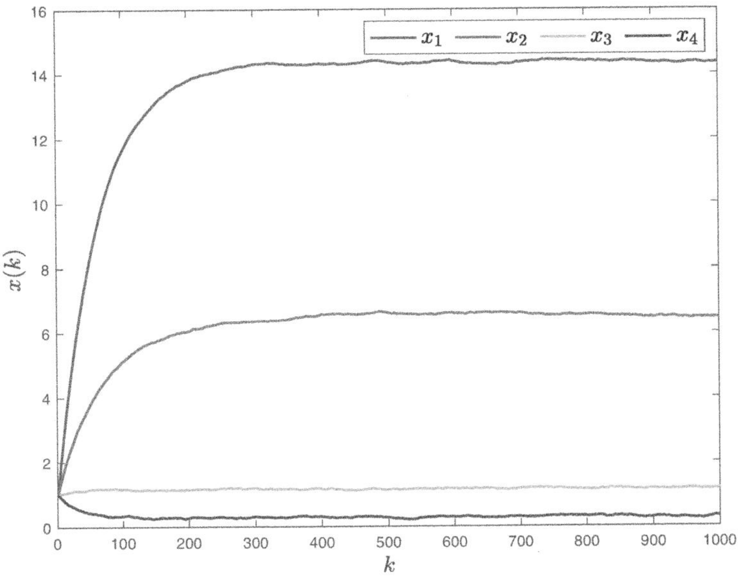

(a) System States

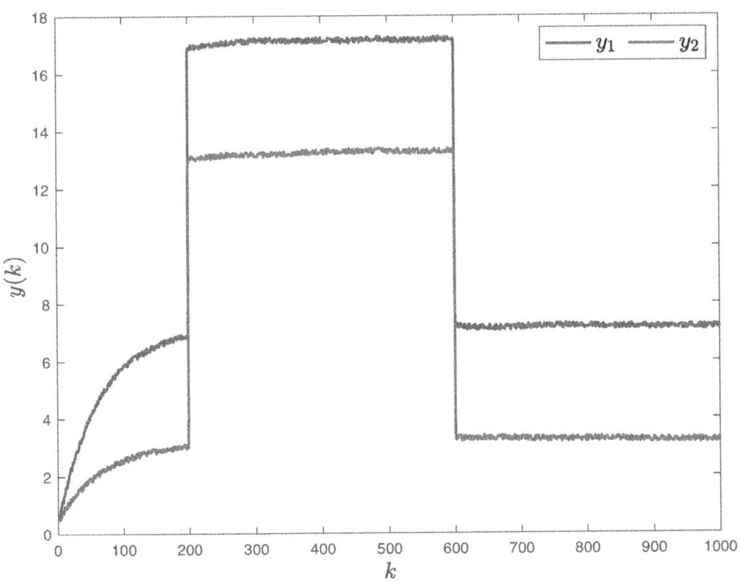

(b) Measurement Outputs

Figure 3.4. *Closed-loop data. For a color version of this figure, see www.iste.co.uk/puig/diagnosis1.zip*

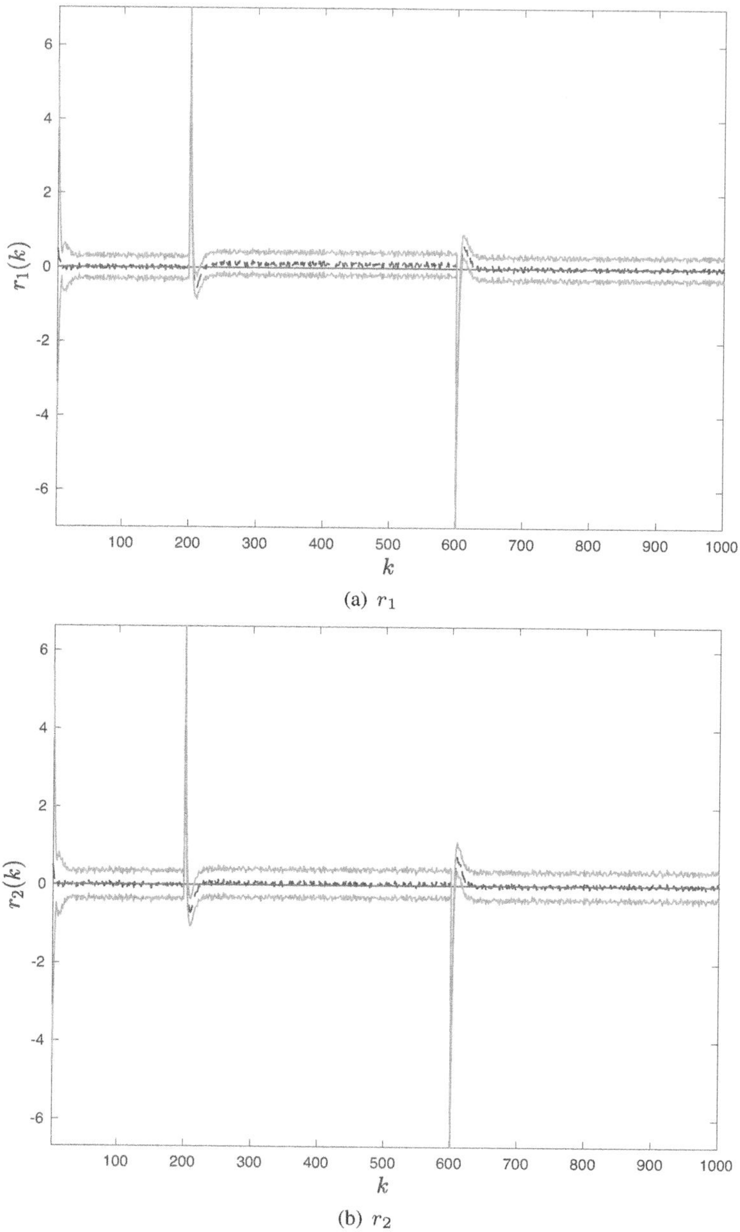

Figure 3.5. *Residual intervals with the set-membership approach. For a color version of this figure, see www.iste.co.uk/puig/diagnosis1.zip*

3.7.1. *Results with robustness condition*

We first use the set-membership approach and the zonotopic Kalman observer to do FD. In this case, only robustness conditions that were introduced in section 3.5.1 are used for the design. In Figure 3.5, the residual intervals with the set-membership approach with the optimal correction matrix are shown. It can be seen that only sensor faults that occurred around $k = 200$ and $k = 600$ can be detected. However, most sensor faults cannot be detected.

In Figure 3.6, the residual intervals with the zonotopic observer and optimal Kalman observer gain are shown. In this case, the effect of uncertainties is minimized. For fault detection, only sensor faults that occurred around $k = 200$ and $k = 600$ can be detected but most sensor faults that occurred still cannot be detected.

3.7.2. *Results with robustness and fault sensitivity conditions*

We next test the design with both robustness and fault sensitivity conditions. In this case, the zonotopic observer structure is chosen. In Figure 3.7, the observer gain design only considers the LMI condition in corollary 3.1. The result is similar to the ones with the zonotopic Kalman observer, where the effect of uncertainties is also minimized. However, most faults cannot be detected.

The observer gain can be designed by considering the robustness condition in corollary 3.1 and the fault sensitivity condition is explained in proposition 3.6. The result is shown in Figure 3.8.

3.8. Conclusion

This chapter has reviewed the use of set-based methods in robust fault diagnosis. Alternatively to the statistical methods, set-membership methods use a deterministic unknown-but-bounded description of noise and parametric uncertainty (interval models). Using approximating sets to enclose the exact set of possible behaviors (in parameter or state space), these methods allow us to check the consistency between observed and predicted behavior. When an inconsistency is detected, a fault can be indicated, otherwise nothing can be stated. The same principle has been used to estimate interval models for fault detection and to develop methods for fault tolerance evaluation. Finally, the application of these methods to the well-known four-tank system case study has been used to exemplify the successful use of the proposed set-membership methods in fault diagnosis.

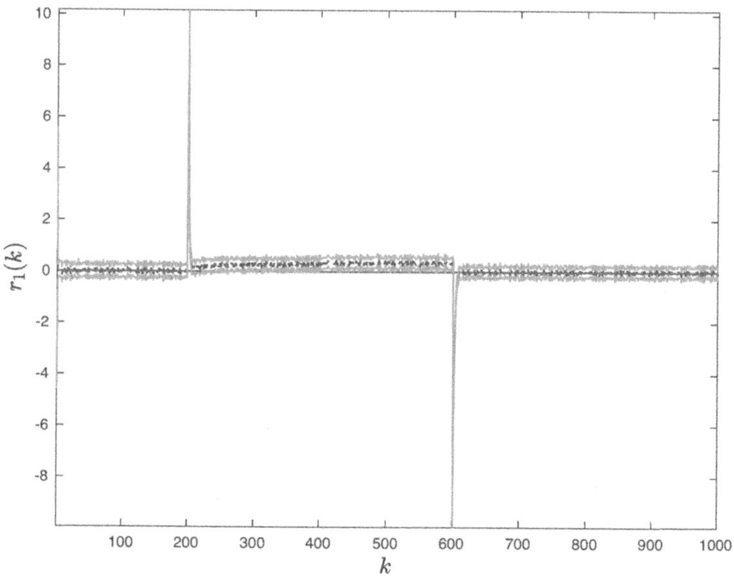

(a) r_1

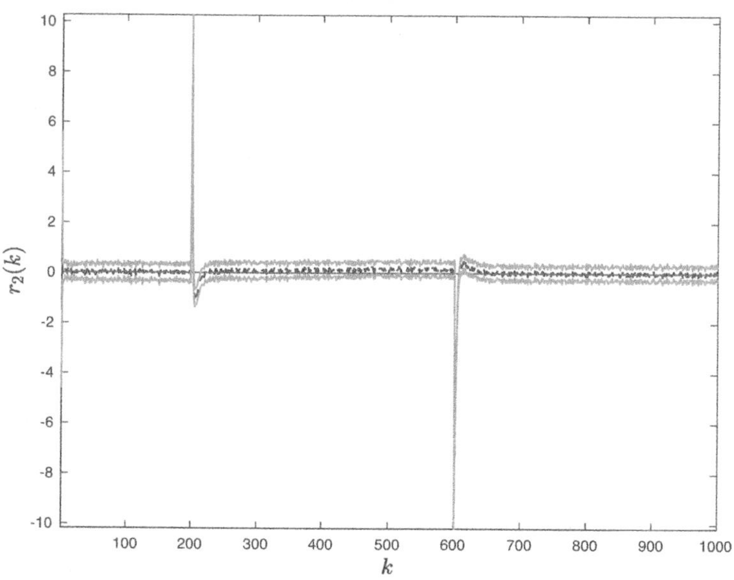

(b) r_2

Figure 3.6. *Residual intervals with zonotopic Kalman observer. For a color version of this figure, see www.iste.co.uk/puig/diagnosis1.zip*

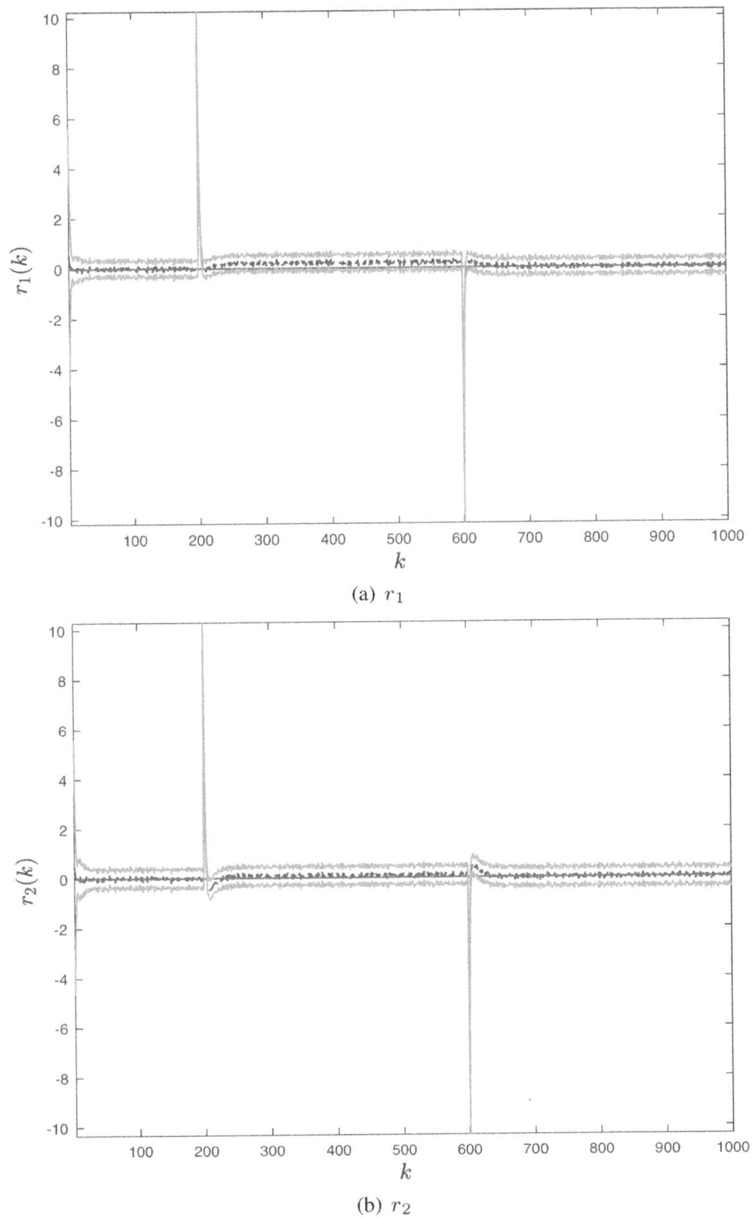

(a) r_1

(b) r_2

Figure 3.7. *Residual intervals with zonotopic observer considering only robustness condition. For a color version of this figure, see www.iste.co.uk/puig/diagnosis1.zip*

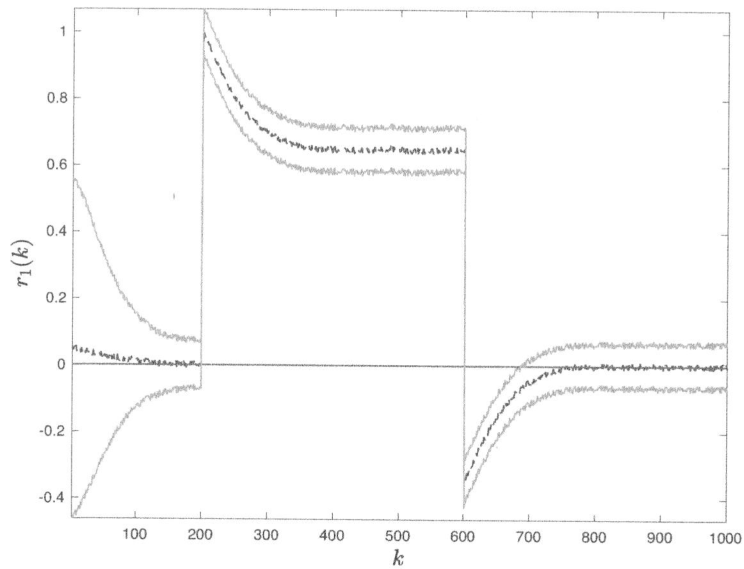

(a) r_1

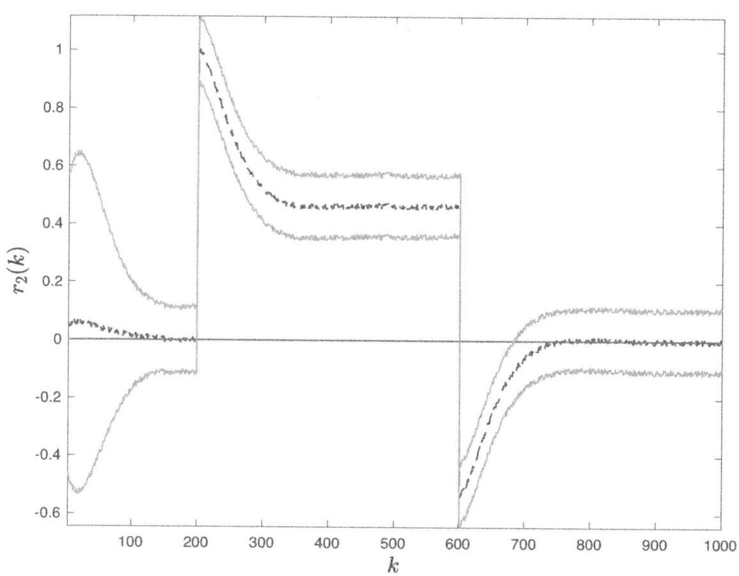

(b) r_2

Figure 3.8. *Residual intervals with zonotopic observer considering both robustness and fault sensitivity conditions. For a color version of this figure, see www.iste.co.uk/puig/diagnosis1.zip*

3.9. References

Ackermann, J. (2002). *Robust Control: The Parameter Space Approach*. Springer, London.

Adrot, O. and Flaus, J.M. (2008). Fault detection based on uncertain models with bounded parameters and bounded parameter variations. In *Proceedings of 17th IFAC World Congress*, 7338–7343, Seoul, South Korea, 6–11 July.

Alamo, T., Bravo, J., Camacho, E. (2005). Guaranteed state estimation by zonotopes. *Automatica*, 41(6), 1035–1043.

Alamo, T., Bravo, J., Redondo, M., Camacho, E. (2008). A set-membership state estimation algorithm based on DC programming. *Automatica*, 44(1), 216–224.

Armengol, J., Veh, J., Sainz, M., Herrero, P., Gelso, E. (2008). Squaltrack: A tool for robust fault detection. *IEEE Transactions on Systems, Man, and Cybernetics: Part B*, 39(2), 475–488.

Bhattacharyya, S., Chapellat, H., Keel, L. (1995). *Robust Control: The Parametric Approach*. Prentice Hall PTR, Englewood Cliffs.

Blanke, M., Kinnaert, M., Lunze, J., Staroswiecki, M. (2016). *Diagnosis and Fault-Tolerant Control*. Springer, Berlin, Heidelberg.

Chen, J. and Patton, R. (1999). *Robust Model-Based Fault Diagnosis for Dynamic Systems*. Kluwer Academic Publishers, Boston.

Combastel, C. (2015). Zonotopes and Kalman observers: Gain optimality under distinct uncertainty paradigms and robust convergence. *Automatica*, 55, 265–273.

Emami-Naeini, A., Akhter, M., Rock, S. (1988). Effect of model uncertainty on failure detection: The threshold selector. *IEEE Transactions on Automatic Control*, 33(12), 1106–1115.

Fagarasan, I., Ploix, S., Gentil, S. (2004). Causal fault detection and isolation based on a set-membership approach. *Automatica*, 40(12), 2099–2110.

Hamelin, F. and Sauter, D. (2000). Robust fault detection in uncertain dynamic systems. *Automatica*, 36(11), 1747–1754.

Horak, D. (1988). Failure detection in dynamic systems with modelling errors. *AIAA Journal of Guidance, Control and Dynamics*, 11(6), 508–516.

Keller, J. and Darouach, M. (1999). Two-stage Kalman estimator with unknown exogenous inputs. *Automatica*, 35(2), 339–342.

Kolev, L. (1993). *Interval Methods for Circuit Analysis*. World Scientific, Singapore.

Le, V.T.H., Stoica, C., Alamo, T., Camacho, E.F., Dumur, D. (2013). *Zonotopes: From Guaranteed State-estimation to Control*. ISTE Ltd, London, and John Wiley & Sons, New York.

Milanese, M., Norton, J., Piet-Lahanier, H., Walter, E. (1996). *Bounding Approaches to System Identification*. Plenum Press, New York.

Ploix, S. and Adrot, O. (2006). Parity relations for linear uncertain dynamic systems. *Automatica*, 42(6), 1553–1562.

Pourasghar, M., Combastel, C., Puig, V., Ocampo-Martinez, C. (2019). FD-ZKF: A zonotopic Kalman filter optimizing fault detection rather than state estimation. *Journal of Process Control*, 73, 89–102.

Puig, V., Cuguero, P., Quevedo, J. (2001). Worst-case state estimation and simulation of uncertain discrete-time systems using zonotopes. *European Control Conference*, IEEE, 1691–1697, Porto, Portugal, 4–7 September.

Puig, V., Stancu, A., Escobet, T., Nejjari, F., Quevedo, J., Patton, R. (2006). Passive robust fault detection using interval observers: Application to the DAMADICS benchmark problem. *Control Engineering Practice*, 14(6), 621–633.

Puig, V., Quevedo, J., Escobet, T., Nejjari, F., De Las Heras, S. (2008). Passive robust fault detection of dynamic processes using interval models. *IEEE Transactions on Control Systems Technology*, 16(5), 1083–1089.

Raïssi, T., Efimov, D., Zolghadri, A. (2012). Interval state estimation for a class of nonlinear systems. *IEEE Transactions on Automatic Control*, 57(1), 260–265.

Rambeaux, F., Hamelin, F., Sauter, D. (2000). Optimal thresholding for robust fault detection of uncertain systems. *International Journal of Robust and Nonlinear Control*, 10(14), 1155–1173.

Sainz, M., Armengol, J., Vehí, J. (2002). Fault detection and isolation of the three-tank system using the modal interval analysis. *Journal of Process Control*, 12(2), 325–338.

Wang, Y., Puig, V., Cembrano, G. (2018). Set-membership approach and Kalman observer based on zonotopes for discrete-time descriptor systems. *Automatica*, 93, 435–443.

4

Diagnosis of Stochastic Systems

Gregory PROVAN
School of Computer Science and IT, University College Cork, Ireland

4.1. Introduction

Addressing uncertainty is a major challenge for accurate fault detection and diagnosis (FDD), given that most system models are inherently uncertain/imprecise (Venkatasubramanian *et al.* 2003). Uncertainty may originate from intrinsic model imprecision or from noisy measurements (e.g. inaccurate calibration due to variation and noise in the data used for model calibration). In addition, specifying the probability for the occurrence of faults is important for two reasons: (1) a point estimate may be inaccurate and (2) omitting the probability implies all faults are equally likely, which can result in inaccurate (multiple-fault) diagnoses in which multiple components may be implicated in the anomalous observation.

The "standard" approach to diagnosis is to define a two-step process of fault identification and isolation, and to provide a method for calculating each in terms of residuals, for example, using analytic redundancy relations (ARRs) (Gertler 1991). In contrast to this algorithmic formulation, we adopt a first-principles framework. This framework is used to classify a collection of stochastic diagnosis methods and provide guidelines on which methods to apply given the task.

We consider the task of diagnosing faults ϕ from limited and imperfect observations $\boldsymbol{y}$. For simplicity, we let both $\boldsymbol{y}$ and ϕ be finite dimensional. We use ϕ to denote the operating modes of a system; these can be faulty or nominal modes.

Diagnosis and Fault-tolerant Control 1,
coordinated by Vicenç PUIG and Silvio SIMANI.

We denote the indirect relationship between observable variables $\boldsymbol{y} \in \mathbb{Y}$ and fault variables $\boldsymbol{\phi} \in \Phi$ using a forward model $\boldsymbol{y} = \xi(\boldsymbol{\phi}, \eta)$, where the forward operator ξ defines how a given fault state gives rise to data, and η is a random variable encompassing measurement noise and/or modeling errors.

We define stochastic diagnosis as an *inverse problem* (Stuart 2010), in that diagnostic inference entails reconstructing unobservable parameters characterizing a system (i.e. the failure state) from (noisy) observations. Inverse problems arise widely in several areas of science and engineering. In more formal terms, we define our inverse problem as reconstructing (estimating) a state variable (or signal) $\phi \in \Phi$, from an observed variable (or data) $\boldsymbol{y} \in \mathbb{Y}$, where Φ and $\mathbb{Y}$ are topological vector spaces.

Two of the most significant challenges of stochastic diagnosis are (1) stochastic model inversion is ill-posed (Vogel 2002), and (2) the computational complexity of the task is high (Stuart 2010). In contrast to the forward inference (mapping inputs to outputs), which is a many-to-one mapping problem, the stochastic inversion problem is one-to-many. More formally, stochastic diagnosis is ill-posed in the following sense: the equation $\boldsymbol{y} = \xi(\boldsymbol{\phi}, \eta)$ is typically a many-to-one mapping function, and hence a diagnosis ϕ is non-unique, especially because unknown noise corrupts the state and measurement equations, and we have limited noisy observations.

Computationally, solving a statistical inverse problem requires many evaluations of the forward model (typically many thousands or even millions), which is not feasible for large-scale systems. Sampling methods for describing probability measures include techniques such as Markov Chain Monte Carlo (MCMC). As a consequence, many approximation methods have been developed, some of which we examine in this chapter.

Using this framework, we can characterize three approaches to stochastic diagnosis as follows (see Figure 4.1):

1) A *model-driven approach* computes a first-principles inverse model to isolate faults, that is $\boldsymbol{\phi} = \xi^{-1}(\boldsymbol{y})$, given the uncertainty η.

2) A *data-driven approach* uses system data $\mathcal{D}$ to compute a pseudo-inverse model to isolate faults, $\boldsymbol{\phi} = \xi_{\Theta}^{-1}(\boldsymbol{y})$, where Θ are the parameters of the learned classifier (pseudo-inverse model) estimated using $\mathcal{D}$.

3) A *hybrid approach* uses system data $\mathcal{D}$ and model ξ to compute an inverse model to isolate faults more efficiently than can a model-driven approach.

Evaluating the effect of uncertainty on a diagnosis algorithm, whether model- or data-driven, is challenging. For example, for model-driven methods like MCMC, assessing the impact of uncertainty requires a large number of samples and running the models with each of these samples may be computationally prohibitive, especially for complex processes (Du *et al.* 2017). Analogously, for data-driven methods this assessment entails significant amounts of data to capture the relevant process behavior, since the calibration of the empirical model is sensitive to the amount and distribution of data; this may be impractical and expensive since some measurements may have little effect on the model calibration.

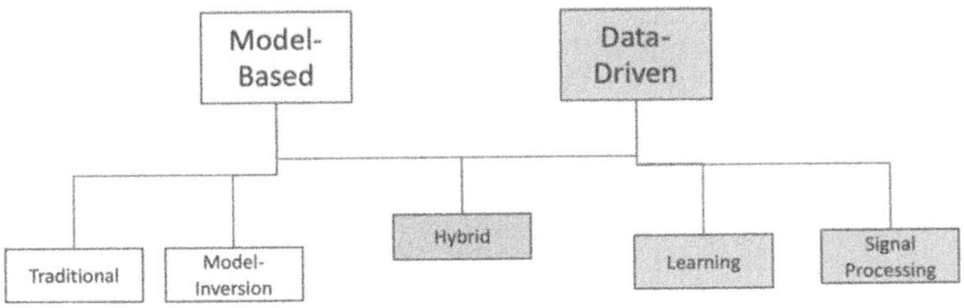

Figure 4.1. *Hierarchy of methods examined. For a color version of this figure, see www.iste.co.uk/puig/diagnosis1.zip*

Rather than compiling the many approaches proposed in the literature, this chapter examines stochastic diagnostic inference from the first-principles point of view of model inversion. We will examine the three main approaches to stochastic diagnostic inference: model-driven, data-driven and hybrid. We will explain the strengths and weaknesses of these approaches, and the prospects for future developments.

This chapter is structured as follows. Section 4.2 introduces our notation and defines our stochastic diagnosis task. Sections 4.4–4.6 introduce model-driven, data-driven and hybrid techniques for stochastic diagnosis, respectively. Section 4.7 compares these approaches, and we conclude the chapter in section 4.8.

4.2. Stochastic diagnosis task

4.2.1. *Notation*

We use the following notation (Table 4.2.1) throughout the chapter.

4.2.2. *Problem formulation*

We define the fundamental relation for a diagnosis model using

$$y = \xi(\phi, \eta), \qquad [4.1]$$

where Φ and $\mathbb{Y}$ are topological vector spaces, $\xi : \Phi \to \mathbb{Y}$ (the forward operator) models how a given signal gives rise to data in the absence of noise, Φ is a Banach/Hilbert space of functions defined on a domain $\Omega \subset \mathbb{R}^d$, $\mathbb{Y}$ is a Banach/Hilbert space of functions defined on a data manifold $\mathbb{M}$ and η is a random variable encompassing measurement noise and/or modeling errors.

$\boldsymbol{x}_{1:k}$	$\{\boldsymbol{x}_1, \boldsymbol{x}_2, \cdots, \boldsymbol{x}_k\}$ vector of states over time
$\boldsymbol{y}_{1:k}$	$\{\boldsymbol{y}_1, \boldsymbol{y}_2, \cdots, \boldsymbol{y}_k\}$ vector of observations over time
$\boldsymbol{\phi}_{1:k}$	$\{\boldsymbol{\phi}_1, \boldsymbol{\phi}_2, \boldsymbol{\phi}_3, \cdots, \boldsymbol{\phi}_k\}$ vector of modes over time
$\boldsymbol{u}_{1:k}$	$\{\boldsymbol{u}_1, \boldsymbol{u}_2, \cdots \boldsymbol{u}_k\}$ vector of control inputs over time
v_n w_n	process noise measurement noise

Table 4.1. *Notation used in the chapter*

Mathematically, we define our diagnosis inverse problem as reconstructing (estimating) a fault variable (or signal) $\phi \in \Phi$, from an observed variable (or data) $\boldsymbol{y} \in \mathbb{Y}$. In physical problems, ξ usually contains a deterministic map that can be decomposed into two elements: the state equations that describe the evolution of the state $\boldsymbol{x}$ in response to the input ϕ, and the output equations that define the relationship between outputs of interest $\boldsymbol{y}$, which in turn define the observations $\boldsymbol{y}$.

DEFINITION 4.1 (Nonlinear System (continuous-time)).– *We model a system Ψ using a set of N interacting variables that define a set of stochastic relations of the form*

$$\dot{\boldsymbol{x}}_t = f(\boldsymbol{x}_t, \boldsymbol{u}_t, \boldsymbol{\phi}_t, v_t), \quad [4.2]$$

$$\boldsymbol{y}_t = g(\boldsymbol{x}_t, \boldsymbol{u}_t, \boldsymbol{\phi}_t, w_t), \quad [4.3]$$

where $\boldsymbol{x}_t$ is the state variable at time t, $\dot{\boldsymbol{x}}_t$ is the derivative of $\boldsymbol{x}_t$ with respect to t, $\boldsymbol{u}_t$ is the vector of inputs (controls), ϕ is the set of modes, $\boldsymbol{y}_t$ is the output vector and v_t and w_t are process and observation noise, respectively.

Where the vector $\boldsymbol{x} \in \mathbb{R}^n$ represents the system states (measured quantities) with initial conditions $\boldsymbol{x}_0 \in \mathbb{R}^n$ over time domain $[0, t_f]$, $\boldsymbol{u}$ denotes the known (measurable) inputs of the process and v is an additive measurement noise vector.

If the governing equations are linear in the state (or if we derive a linearized model of [4.2] and [4.3] by linearizing about an equilibrium point), then the system can be rewritten as

$$\dot{\boldsymbol{x}} = A_{\boldsymbol{\phi}}\boldsymbol{x} + B_{\boldsymbol{\phi}}\boldsymbol{u} + \gamma(\phi, t), \qquad [4.4]$$

$$\boldsymbol{y} = C_{\boldsymbol{\phi}}\boldsymbol{x}, \qquad [4.5]$$

where $A,\ B \in \mathbb{R}^{N\times N}$ is a matrix that possibly depends on the parameters but not on the state, and the general nonlinear function $\gamma \in \mathbb{R}^N$ represents the direct contributions of the parameters to the governing equations, as well as forcing due to boundary conditions and source terms. For the linearized output equation, $C \in \mathbb{R}^{q\times N}$ is a matrix that maps states to outputs, and we have assumed no direct dependence of outputs on parameters (without loss of generality, such a term could easily be incorporated).

4.2.3. *Representing uncertainty*

There are several ways to represent uncertainty in a system model, and we address the following two main methods: (1) additive noise and (2) intrinsic uncertainty. The *additive noise approach* represents a system as $y = \xi(\phi) + \delta_y$, where δ_y is the noise added to a deterministic operator ξ. The *intrinsic uncertainty approach* represents a system in terms of a probability distribution $y = P(y|\phi)$. Sharifian *et al.* (2019) present a survey of the general issues for uncertainty in chemical process systems engineering, and how it is encoded.

The standard *intrinsic uncertainty approach* is to use a probabilistic formulation, in which the likelihood function $\pi(\boldsymbol{y}|\boldsymbol{\phi})$ describes the probability of measurements $\boldsymbol{y}$ given a particular value of $\boldsymbol{\phi}$. The likelihood function thus incorporates the forward model and the measurement noise, and results from the specification of ξ. A stochastic formulation of an inverse task makes its solution even more challenging (Stuart 2010).

If we adopt a Bayesian formulation of the inverse problem, we encode $\boldsymbol{\phi}$ as a random variable, whose prior probability density $\pi_{pr}(\boldsymbol{\phi})$ encodes prior knowledge about the inputs. By applying Bayes' rule, we obtain the posterior probability density of the inputs, π_{post}, as follows:

$$\pi_{post}(\phi) \equiv P(\boldsymbol{\phi}|\boldsymbol{y}) \propto P(\boldsymbol{y}|\boldsymbol{\phi})\pi_{pr}(\boldsymbol{\phi}). \qquad [4.6]$$

Equation [4.6] thus defines the solution of the inverse problem as a probability density for the model inputs $\boldsymbol{\phi}$.

In most practical situations and in the Bayesian formulation described in equation [4.6], the observational data are finite-dimensional on some domain D; thus $\boldsymbol{y} \in$

$D \subseteq \mathbb{R}^M$ consists of $\boldsymbol{y}$ evaluated at a finite set of times, $\boldsymbol{y} = (\boldsymbol{y}(t_1), ..., \boldsymbol{y}(t_{n_t}))$ with $M = n_t q$. We denote the deterministic forward model mapping inputs $\boldsymbol{\phi}$ as finite-time observations $\boldsymbol{y}$ by $\xi(\boldsymbol{\phi}) : \Phi \rightarrow D$. With nonlinear forward models, the inverse solution is typically represented by samples simulated from the posterior distribution (equation [4.6]). From these samples, posterior moments, marginal distributions and other summaries can be evaluated. Solution of the statistical inverse problem thus requires many evaluations of the forward model (typically many thousands or even millions); this is a computationally prohibitive proposition for large-scale systems.

4.3. Inference methods for diagnosis task

This section provides an overview of the inference methods for computing diagnoses. Diagnosis at time t, corresponding to an anomalous observation $\boldsymbol{y}_t$, is a model-inversion task that computes the fault-state, ϕ_τ, $\tau \leq t$. In other words, if we have a (simplified) system representation $\boldsymbol{y} = \xi(\boldsymbol{\phi})$, diagnosis inference consists of computing the inverse model $\boldsymbol{\phi} = \xi^{-1}(\boldsymbol{y})$.

We can define three different approaches to solving this diagnosis task:

– A **model-based approach** inverts the system model in order to compute ϕ.

– A **data-driven approach** estimates a parameterized pseudo-inverse directly from data, that is, it estimates a diagnosis classifier $\boldsymbol{\phi} = \xi_\Theta^{-1}(\boldsymbol{y})$ with parameters Θ.

– A **hybrid approach** integrates elements of model-based and data-driven approaches.

4.3.1. *Difference with other tasks*

Diagnosis is similar to but different from other well-known tasks. We clarify the difference with two tasks below.

The **state estimation task** is to compute an estimate of $\boldsymbol{x}_k$ using $\mathbb{Y}_k$ at each time k. This is accomplished by computing $P(\boldsymbol{x}_k|\mathbb{Y}_k)$, that is, the probability density function (PDF) of $\boldsymbol{x}_k$ conditioned on $\mathbb{Y}_k$. This is because $P(\boldsymbol{x}_k|\mathbb{Y}_k))$ captures the information of $\boldsymbol{x}_k$ conveyed in $\mathbb{Y}_k$ and can be leveraged to estimate $\boldsymbol{x}_k$. Rather than reconstructing the entire system state, we are interested in just the fault state, that is $P(\boldsymbol{\phi}_k|\mathbb{Y}_k)$. Further, many state estimation approaches assume a fault-free system.

The **state forecast task** is to compute an estimate of $\boldsymbol{x}_{k+1}$ using $\boldsymbol{x}_k$ and $\mathbb{Y}_k$. This is accomplished by computing $P(\boldsymbol{x}_{k+1}|\mathbb{Y}_k, \boldsymbol{x}_k)$.

4.4. Model-based approach

First-principles model-based methods use physical laws for describing the dynamic behavior of the monitored system to generate a model, and then these methods use an algorithm that inverts that model for diagnostics inference. This section describes the main model types and their corresponding inversion algorithms. We group the approaches in terms of (1) traditional FDD methods and (2) Bayesian inversion methods.

An abstract view of model-based inversion is that we aim to minimize the misfit against data $\boldsymbol{y}$. For example, we can minimize a loss function $\mathcal{L}(\xi(\phi, \boldsymbol{y}))$, where $\xi\colon Y \times Y \to \mathbb{R}$ is a suitable affine transformation of the data log-likelihood, which is equivalent to finding a maximum likelihood solution.

4.4.1. *Traditional FDD methods*

4.4.1.1. *Overview of FDD methods*

Traditional FDD algorithms are based on residual reasoning. Residuals are the difference between observed behavior of a system (defined by the actual measurement $\boldsymbol{y}$) and the predicted measurement $\hat{\boldsymbol{y}}$, which denotes the forecast behavior of the system, as derived from the dynamical system model. Generating the residual, and then isolating a fault, is based on a model inversion step that transforms the dynamical equations into a representation that facilitates this inference.

FDD methods typically involve two steps: (i) residual generation obtained from the differences of model predictions and data, and (ii) fault isolation evaluating the residual for decision-making. We define these steps as follows.

Fault detection: compute $\boldsymbol{r}$, the residual vector, which is given by

$$\boldsymbol{r}_t = \| \boldsymbol{y}_t - \hat{\boldsymbol{y}}_t \|, \qquad [4.7]$$

where $\hat{\boldsymbol{y}}_t$ corresponds to the vector of estimated outputs and $\| \cdot \|$ is an appropriate norm.

To provide useful information for fault diagnosis, a residual for a deterministic system should be defined in such a way that

$$\boldsymbol{r}_t \mapsto \begin{cases} \simeq 0 & \text{if fault-free} \\ \neq 0 & \text{if fault exists} \end{cases} \qquad [4.8]$$

A residual detects any changes in the mean values of failure modes, $\{\bar{\phi}_i\}$, $i = 1, \cdots, m$, in the presence of stochastic perturbations. The residual is often generated with either an observer for deterministic models (Patton and Chen 1997) or a filter for a stochastic process (Alrowaie *et al.* 2012). These methods typically use an additive noise model.

For stochastic systems, we define residuals in terms of density functions. In this case, a “zero residual" (indicating a nominal system) corresponds to the case where the density function of the residual is statistically equivalent to that of the system’s measurement noise; the residual is “non-zero” otherwise. We can use one of many statistical techniques to evaluate the residuals for stochastic systems, such as the log-likelihood ratio (LLR), the generalized log-likelihood ratio (GLR), the sequential probability ratio test (SPRT) or the cumulative sum test statistic (CUSUM). For nonlinear systems, straightforward residual analysis is not possible and therefore more advanced statistical tests must be used (see, e.g., Alrowaie *et al.* (2012)).

Fault isolation: isolate the fault, that is, the specific mean value $\bar{\phi}_i$ at a given time instant t. This phase of inference typically uses statistical testing to identify the faults.

Fault isolation algorithms differ based on the underlying inversion method and residual definition. We now review two well-known approaches for FDD, analytical redundancy relations (ARR) and observer-based methods.

4.4.1.2. *Examples of FDD methods*

Analytical redundancy relations/structural methods: this class of method reformulates the equations for improved diagnostics inference. There are two closely related approaches, namely analytical redundancy relations (ARR) – also called residual-generation methods (Gertler 1991) – and parameter estimation. Given a set of dynamical equations ξ, this class of algorithm transforms ξ into a new representation ξ' that (a) identifies a subset of the system in which faults can be computed (the diagnosable subset) and (b) facilitates residual generation and evaluation.

Let us consider a linear system. We can define a diagnosable subset of variables as those for which there are at least as many equations as variables (that is, based on the fully determined and overdetermined sets of equations); the remaining subset of variables is not diagnosable and is referred to as underdetermined. ARRs can be used for diagnosis only of diagnosable sub-systems. In other words, evaluating an ARR is possible only for overdetermined and fully determined subsets of equations (Staroswiecki and Declerck 1990).

An ARR (Cordier *et al.* 2004) is a constraint deduced from the model of the system containing only measured variables, thereby facilitating residual analysis from any observation. The ARRs are satisfied if the observed system behavior satisfies the model constraints, which is equivalent to a zero residual.

ARRs also facilitate fault isolation. The ARRs relate measured deviations in the variables to the diagnosable failure-mode variables of the system. Fault isolation, that is, residual evaluation, is thus defined in terms of a logical combination of measurement deviations and implicated fault parameters.

We compute ARRs from the set of overdetermined equations obtained from the structural system model using an elimination procedure, e.g. using structural analysis based on Mendelsohn–Dulmage model reconstruction (Staroswiecki and Declerck 1990). A structural system description specifies a matrix denoting which variables participate in which equations. This procedure then computes a unique canonical decomposition of the structural description of the system, identifying the under-determined, just-determined and overdetermined sets of equations. The canonical decomposition is based on finding a complete matching, with respect to unknown variables, in the bipartite graph associated with the structural description of the system. A combination of just-determined systems together with redundant relations is the basis for an ARR. Faults are isolated using the set of overdetermined systems.

Observer-based methods: the observer-based approach uses a parametric design technique to perform both detection and diagnosis. Given a linear forward model described in terms of matrix-based mappings (equations [4.4] and [4.5]), we compute an inverse model in terms of a matrix-based transformation, assuming that the process and observation matrices satisfy appropriate conditions. Few results in the literature consider the fault detection and estimation problem for stochastic systems; (see, e.g., Chen and Patton (1996)) for an early formulation.

The residual is computed using an observer relation, as shown in Figure 4.2. The observer consists of a model parallel to the process with feedback of the output estimation error, $\boldsymbol{r}_t = \boldsymbol{y}_t - \hat{\boldsymbol{y}}_t$. Variable $\hat{\boldsymbol{y}}_t$ corresponds to the vector of estimated outputs, and K is the observer gain matrix.

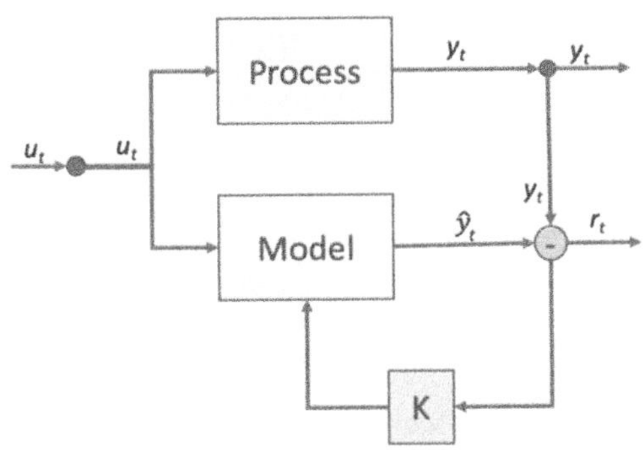

Figure 4.2. *Architecture of observer-based approach. For a color version of this figure, see www.iste.co.uk/puig/diagnosis1.zip*

The above observer model works for a single fault only; to diagnose multiple simultaneous faults, a bank of observers is required, one for each possible fault.

4.4.2. *Bayesian inversion/filtering*

A Bayesian formulation of the inverse problem explicitly encodes probabilities into the dynamics and the prior beliefs over the state variables. We can use a discrete-time formulation of this as follows:

DEFINITION 4.2 (Stochastic System (discrete-time)).–

$$\boldsymbol{x}_{k+1} : P(\boldsymbol{x}_{k+1}|\boldsymbol{x}_k, \boldsymbol{u}_k, \boldsymbol{\phi}_k), \qquad [4.9]$$

$$\boldsymbol{y}_k : P(\boldsymbol{y}_k|\boldsymbol{x}_k, \boldsymbol{u}_k, \boldsymbol{\phi}_k). \qquad [4.10]$$

Equation [4.9] characterizes the state transition probability, and equation [4.10] describes the stochastic measurement model.

Given the model described by equations [4.9] and [4.10], our stochastic diagnosis problem consists of computing the health state $P(\boldsymbol{\phi}_k|\boldsymbol{y}_{1:k})$ given that we have observed $\boldsymbol{y}_{1:k}$. To perform this inference, we use Bayes' rule, which yields:

$$P(\boldsymbol{\phi}_k|\boldsymbol{y}_{1:k}) = \frac{P(\boldsymbol{\phi}_k, \boldsymbol{y}_{1:k})}{P(\boldsymbol{y}_{1:k})} = \frac{P(\boldsymbol{y}_{1:k}|\boldsymbol{\phi}_k)P(\boldsymbol{\phi}_k)}{P(\boldsymbol{y}_{1:k})}. \qquad [4.11]$$

We now describe this in terms of *stochastic filtering*, which solves a statistical inversion problem. We estimate an unknown vector-valued time series $\boldsymbol{\phi}_{1:k} \equiv \{\boldsymbol{\phi}_1, \boldsymbol{\phi}_2, \boldsymbol{\phi}_3, \cdots\}$, which is observed through a set of noisy measurements $\boldsymbol{y}_{1:k} \equiv \{\boldsymbol{y}_1, \boldsymbol{y}_2, \cdots\}$. In the following, we simplify our notation for the purposes of exposition, suppressing $\boldsymbol{u}$, v_n and w_n.

DEFINITION 4.3 (Statistical inversion problem).– *Statistical inversion estimates the hidden states $\boldsymbol{\phi}_{1:k}$ from the observed measurements $\boldsymbol{y}_{1:k}$. To do this, we use Bayes' rule to compute the joint posterior distribution of the states given the measurements as:*

$$P(\boldsymbol{\phi}_{1:k}|\boldsymbol{y}_{1:k}) \propto \frac{P(\boldsymbol{\phi}_{1:k})P(\boldsymbol{y}_{1:k}|\boldsymbol{\phi}_{1:k})}{P(\boldsymbol{y}_{1:k})}. \qquad [4.12]$$

where

– $P(\boldsymbol{\phi}_{1:k})$ *is the prior distribution defined by the dynamic model;*

– $P(\boldsymbol{y}_{1:k}|\boldsymbol{\phi}_{1:k})$ *is the likelihood model for the measurements;*

– $P(\boldsymbol{y}_{1:k})$ *is the normalization constant defined as*

$$P(\boldsymbol{y}_{1:k}) = \int P(\boldsymbol{y}_{1:k}|\boldsymbol{\phi}_{1:k})P(\boldsymbol{\phi}_{1:k})d\boldsymbol{\phi}_{1:k} \qquad [4.13]$$

Although the posterior density provides a complete solution of the stochastic filtering problem, the problem still remains intractable since the density is a function rather than a finite-dimensional point estimate.

Unfortunately, computing the full posterior formulation is intractable, and it must be recomputed each time we obtain a new measurement. This is further complicated by the fact that with each new time step, the computational complexity of that step increases due to the increased dimensionality of the full posterior distribution. Hence, additional information or restrictive approximations on the full posterior computation (or other modeling assumptions) are necessary to achieve computational tractability.

4.4.2.1. *Inference methods*

We can apply a variety of inference algorithms for this Bayesian stochastic inversion task. These include exact methods, sampling-based methods and variational methods (Blei *et al.* 2017), all of which we briefly review.

In general, it is hard to obtain information from a probability measure in high dimensions. One useful approach to extracting information is to find a maximum *a posteriori* estimator: a point u which maximizes the posterior PDF; such variational methods are surveyed later in the chapter.

Exact methods: these apply an inference algorithm to compute the posterior $P(\boldsymbol{\phi}|\boldsymbol{y})$ directly. For example, if we encode the probability model as a graphical model $\mathcal{G}(Z)$, where $Z = \boldsymbol{\phi} \cup \boldsymbol{y}$, then we can use general-purpose algorithms, such as join-tree decomposition and propagation (Koller *et al.* 2009).

Alternatively, if the graphical model takes a particular form, we can use an algorithm purposely designed for that form. Examples of application-specific graphical models include hidden Markov models (Smyth 1994), Kalman filters (Kalman 1960), particle filters (Doucet and Johansen 2009) and dynamic Bayes networks (Koller *et al.* 2009). Provan (2019) presents an overview of diagnostic inference with such probabilistic graphical models.

Instead of computing an exact solution, we can instead compute a solution with a bounded error, that is, a robust diagnosis. One approach, robust inversion (see El Amri *et al.* (2020)), computes the set of observed variables $\boldsymbol{y} \in \mathbb{Y}$ such that $\sup_{\boldsymbol{\phi}\in\Phi} \xi(\boldsymbol{y}, \boldsymbol{\phi}) \leq c$ for some threshold c.

Sampling-based methods: another commonly used method for computing a probability measure in high dimensions is *sampling*. This involves generating a set of samples $\{\phi_i|\boldsymbol{y}\}_{i=1}^{N}$ distributed (perhaps only approximately) according to $P(\phi|\boldsymbol{y})$. One commonly-used sampling method is Markov chain Monte Carlo (MCMC) (Berg and Billoire 2007). MCMC sampling operates in three steps: (1) construct an ergodic Markov chain on ϕ, whose stationary distribution is the posterior $P(\phi|\boldsymbol{y})$; (2) sample from the chain to collect samples from the stationary distribution; and (3) approximate the posterior with an empirical estimate constructed from the collected samples. A drawback of MCMC is that a large number of samples are needed to accurately approximate a particular distribution for a large number of variables.

Variational methods: variational inference solves the inference problem as an optimization problem, using a family $\mathcal{F}$ of approximate densities to control the complexity of this optimization such that we can generate a density that is both close to $P(\phi|\boldsymbol{y})$ and also simple enough for efficient optimization. The standard way to do this is to compute the member of $\mathcal{F}$ that minimizes the Kullback–Leibler (KL) divergence to the exact posterior:

$$\pi^*(\phi) = \arg \min_{\pi(\phi)\in\mathcal{F}} KL[\pi(\phi) \parallel P(\phi|\boldsymbol{y})]. \qquad [4.14]$$

We then use this to approximate the posterior with the optimized member of the family $\pi^*(\cdot)$.

Variational methods simplify the inference task by decoupling the degrees of freedom in the original problem; they expand the problem to include additional parameters, called variational parameters, that must be fit for the problem at hand.

Compared to MCMC, variational inference tends to be faster and scales well to large data; however, it is less well understood (from both theoretical and empirical perspectives) than MCMC.

4.5. Data-driven approaches

This section describes two common data-driven approaches based on machine learning (ML) and statistical analysis. Data-driven models are well suited to cases where the underlying physical process is unknown or is too complicated to be described. Training suitable models requires data that reflect the underlying process.

The success of data-driven diagnosis depends on the amount and "quality" of data available for training. By quality, we mean the distribution of nominal and faulty data; with overly sparse faulty data, it is difficult to train classifiers to recognize those faults. To address this issue, researchers often use a class of models that just identifies a system as being normal or abnormal. This simple binary classifier only requires normal historical data, from which models are easily and quickly built.

However, the major drawback of data-driven models (like multivariate statistical approaches) is explainability; unlike model-based approaches like parity equation approaches, they do not provide a causal description of the process that can be used as the basis for fault isolation (diagnosis).

4.5.1. *ML methods*

ML algorithms are usually classified into three categories: supervised, semi-supervised and unsupervised:

– *supervised ML* assumes that *all data* are labeled according to fault status and this requires data for nominal/failures modes and their expected behaviors;

– *semi-supervised ML* assumes that *some data* are labeled according to fault status and this requires less complete data, e.g. often only healthy data are available;

– *unsupervised ML* assumes that *no data* are labeled according to fault status and this approach is limited to clustering data according to common observations, for example, vibration signatures.

In the following, we focus on the supervised case. For more thorough reviews of ML-based diagnostics, we refer the reader to various studies (Khan and Yairi 2018; Zhao *et al.* 2019; Ademujimi *et al.* 2017; Lo *et al.* 2019; Rezaeianjouybari and Shang 2020). Nor *et al.* (2020) review the data-driven methods in FDD systems for chemical process systems.

ML can create a nonlinear function approximation for the inverse problem in equation [4.1]; we state this as the problem of finding a (nonlinear) mapping $\xi_\Theta^{-1} : \mathbb{Y} \to \Phi$ satisfying the following pseudo-inverse property:

$$\xi_\Theta^{-1} \simeq \phi \quad \text{whenever equation [4.1] holds} \qquad [4.15]$$

A key element is to parameterize the set of such pseudo-inverse operators by $\Theta \in \mathbb{Z}$, where $\mathbb{Z}$ is a suitable parameter space. The parameters in data-driven models adjust and control all underlying processes and relationships. Learning identifies an "optimal" $\boldsymbol{\Theta}$ given training data, where the concept of optimality is typically quantified through a loss functional that measures the "quality" of a learned pseudo-inverse ξ_Θ^{-1}. The loss functional must be specified based on the type of training data, that is, we define supervised or unsupervised learning. Researchers have developed diagnostics methods based on ML approaches such as artificial neural networks, neuro-fuzzy methods, support vector machines, Gaussian mixture model, K-nearest neighbor and Bayesian networks. Most ML applications to stochastic diagnosis have been limited to single component systems, such as condition monitoring for motors or drives (Zhao *et al.* 2019; Zhang *et al.* 2019). Few system-level learning applications have been developed due to the challenge of compiling sufficient system-level data, including data with fault conditions.

In contrast to model-based methods, in which component-level models can be integrated into a system-level model, the black-box nature of ML models does not enable compositional development of system-level models, since there is no clear notion of state variables necessary to develop system-level representations.

4.5.2. *Statistical methods*

Many different multivariate statistical analysis methods for diagnostics have been proposed in the literature such as principal component analysis (PCA), partial least squares, independent component analysis and Fisher discriminant analysis. These approaches extract variable correlations from data to reduce dimensionality and to identify faults in a process (Chiang *et al.* 2000). Multivariate statistical classifiers easily handle very large problems involving hundreds of variables, are insensitive to noise in the measured variables and, very importantly, handle missing data in an efficient and easy manner.

These approaches perform a data-driven inversion in which the data are transformed into an alternative representation from which statistical testing can more readily isolate faults. In contrast to ML approaches, which create classifiers that map data to failure mode, these statistical approaches typically require a statistical test based on the transformed data.

These approaches isolate faults using a two-step process:

1) They transform the data into a representation to simplify mode estimation. For example, PCA and projections to latent structures (PLS) project most of the important information in these data down into low dimensional latent variable spaces, where regions of normal and abnormal operation are easily defined.

2) They use statistical tests to detect any abnormal behavior in new data. For example, the principal components score plot (T2) and squared prediction error (SPE) plot can be employed. The most common fault isolation method is the contribution plot, that is, plots of the contributions of each variable to the deviation of the current operating point from that defined by the PCA or PLS model for normal in-control operation. These plots are usually very powerful at isolating simple sensor or actuator faults but much less successful at clearly isolating complex faults.

A major deficiency of these multivariate statistical approaches is that the existing diagnosis algorithms, such as contribution plots, still have some trouble when they are applied in real industrial processes. In addition, unlike the conventional state variable model or parity equation approaches, they do not provide a causal description of the process that can be used as the basis for fault isolation (diagnosis). These statistical models also suffer from the same limitation of ML-based models, i.e. limitation to single component systems and inability to compose larger systems from single component models.

4.6. Hybrid approaches: surrogate methods

Hybrid approaches (also known as gray-box approaches) combine model- and data-driven approaches (Bohlin 2006; Mattis and Wohlmuth 2018; Xiong and Jutan 2002). Surrogate models address a well-known challenge of stochastic inversion, that is, its extreme computational costs, by using simplified and/or hybrid models. McBride and Sundmacher (2019) provide an overview of surrogate modeling in chemical process engineering and Bhosekar and Ierapetritou (2018) provide an overview for model design optimization.

Surrogate methods replace some portion of a model with a data-driven representation (surrogate model) for increased computational efficiency or due to an inability to properly model some aspects of the underlying system. Although the model is not completely described by physical equations, the combination of equations and parameters remains physically interpretable.

Surrogate models are designed to approximate the dominant features of a complex model (Bhosekar and Ierapetritou 2018). Two common approaches for constructing surrogate models are as follows: (1) use sampling to generate a model ξ_s that is simpler than the full model ξ and (2) use data $\mathcal{D}$ together with a simplified model to generate a data-augmented model $\xi_{s,\theta}$ that better approximates the full model than the simplified model alone.

We can frame the surrogate-generation process as follows: we aim to generate an analytical or numerical surrogate model ξ_s to replace a (high-fidelity) model $\xi(\phi)$ in a diagnosis task.

4.6.1. *Fitting surrogate models via sampling*

We can define surrogate-generation via sampling as follows (Garud *et al.* 2017; Mattis and Wohlmuth 2018):

Given:

– $\boldsymbol{y} = \xi(\phi)$ and

– a mathematical form for ξ_s,

compute the sampling set $\mathcal{S}$ that gives ξ_s with a prescribed accuracy for approximating $y = \xi(\phi)$ for $\phi \in \Phi$.

4.6.1.1. *Fitting surrogate models with data*

A second approach to surrogate model design is to tune a simpler model with data or to replace part of the model with a learned representation.

Hence, our problem is as follows.

Given:

– $y = \xi(\phi)$,

– a collection of low-fidelity models $\tilde{\Xi} = \{\tilde{\xi}_1(\phi), \cdots, \tilde{\xi}_q(\phi)\}$,

– data $\mathcal{D}$ and

– a mathematical form for ξ_s,

compute the representation ζ such that $\zeta(\tilde{\xi}_1(\phi), \cdots, \tilde{\xi}_q(\phi))$ approximates $\xi(\phi)$ (for $\phi \in \Phi$) with a prescribed accuracy.

There are two approaches for data-driven surrogate diagnostics:

Data-driven tuning: this approach uses data $\mathcal{D}$ to improve the performance of a simplified model $\tilde{\xi}$, or simplified ensemble of models $\tilde{\xi}_i$, $i = 1, ..., q$. For example, Provan (2017) uses a polynomial regression approach for tuning the diagnostic performance of an ensemble of low-fidelity ODE diagnosis models such that we achieve the target of embedded processing limits.

Data-driven augmentation: this approach learns a portion of a model from data $\mathcal{D}$ to augment a set of equations. For example, Yang and Perdikaris (2019) use deep learning to generate multi-fidelity models of stochastic processes based on different information sources of variable fidelity. Another deep learning approach for Bayesian inversion is Yan and Zhou (2019).

4.7. Comparative analysis of approaches

Even though a wide variety of approaches have been developed for stochastic diagnosis, adopting a model-inversion perspective enables a clear comparison of these methods. All methods aim to transform observed data into a diagnosis (an assignment of a posterior probability distribution to failure mode variables ϕ) through a process of model inversion.

Model-driven approaches directly use the dynamical equations specifying the forward model, $y = \xi(\phi, \eta)$. The FDD methods compute a pseudo-inverse of ξ and then perform a statistical test for fault isolation. Bayesian filtering methods do not require statistical testing but compute the posterior fault distribution directly for an inversion algorithm.

Data-driven approaches do not use a manually-generated model but induce a pseudo-inverse model ξ_{Θ}^{-1} (parameterized by Θ) from data. Analogous to the model-driven approaches, data-driven FDD methods compute a pseudo-inverse of ξ and then perform a statistical test for fault isolation. In contrast, ML approaches induce a classifier that enables direct diagnostics given an observation, using $\phi = \xi_{\Theta}^{-1}(\boldsymbol{y})$.

Hybrid approaches attempt to use a computationally simpler model than model-driven approaches using data to tune the process.

All of these methods have their strengths and weaknesses. *Model-driven approaches* are easily interpreted and have good extensibility for fault isolation. In addition, model libraries enable large systems to be composed of component models. However, the models are laborious to construct manually, and inference is expensive.

Data-driven approaches avoid the need for laborious manual model construction; however, learning models requires adequate to large amounts of data with suitable distributions of nominal and fault conditions. For this class of approach, diagnostics inference is typically inexpensive. However, these methods are not easily interpreted, do not have good extensibility for fault isolation and cannot use model libraries to develop large systems from component models.

Hybrid approaches have the potential for enabling the benefits of both model- and data-driven methods, that is, models that are computationally simpler than model-driven approaches and also require less manual model development. However, significant research is necessary to develop this approach for real-world diagnostics applications.

4.8. Summary and conclusions

This chapter has surveyed methods for diagnosing stochastic systems from a first-principles perspective. We have put a large family of heterogeneous algorithms into a consistent framework, thereby enabling easy comparison of strengths and weaknesses of the approaches. We have also introduced an emerging family of methods that are a hybrid of the well-studied model- and data-driven approaches.

Future prospects: stochastic diagnosis is benefitting from the increase in data and learning algorithms, particularly deep learning algorithms. However, advances are necessary to enable deep learning to provide meaningful contributions to the diagnosis of large-scale systems. At present, deep learning has proven beneficial only to component-based diagnostics and it is unclear how to scale this up to large systems. These drawbacks may be avoided through the use of hybrid methods that can blend the model structure for large systems with local fault classifiers.

4.9. References

Ademujimi, T.T., Brundage, M.P., Prabhu, V.V. (2017). A review of current machine learning techniques used in manufacturing diagnosis. In *Advances in Production Management Systems: The Path to Intelligent, Collaborative and Sustainable Manufacturing: IFIP WG 5.7 International Conference*, Riedel, R., Thoben, K-D., Kiritsis, D., von Cieminski, G., and Lodding, H. (eds), Springer, New York.

Alrowaie, F., Gopaluni, R., Kwok, K. (2012). Fault detection and isolation in stochastic non-linear state-space models using particle filters. *Control Engineering Practice*, 20(10), 1016–1032.

Berg, B.A. and Billoire, A. (2007). Markov chain Monte Carlo simulations. In *Wiley Encyclopedia of Computer Science and Engineering*, John Wiley & Sons, Inc, Hoboken.

Bhosekar, A. and Ierapetritou, M. (2018). Advances in surrogate based modeling, feasibility analysis, and optimization: A review. *Computers & Chemical Engineering*, 108, 250–267.

Blei, D.M., Kucukelbir, A., McAuliffe, J.D. (2017). Variational inference: A review for statisticians. *Journal of the American Statistical Association*, 112(518), 859–877.

Bohlin, T.P. (2006). *Practical Grey-box Process Identification: Theory and Applications*. Springer Science & Business Media, London.

Chen, J. and Patton, R.J. (1996). Optimal filtering and robust fault diagnosis of stochastic systems with unknown disturbances. *IEE Proceedings – Control Theory and Applications*, 143(1), 31–36.

Chiang, L.H., Russell, E.L., Braatz, R.D. (2000). *Fault Detection and Diagnosis in Industrial Systems*. Springer Science & Business Media, London.

Cordier, M.-O., Dague, P., Lévy, F., Montmain, J., Staroswiecki, M., Travé-Massuyès, L. (2004). Conflicts versus analytical redundancy relations: A comparative analysis of the model based diagnosis approach from the artificial intelligence and automatic control perspectives. *IEEE Transactions on Systems, Man, and Cybernetics – Part B: Cybernetics*, 34(5), 2163–2177.

Doucet, A. and Johansen, A.M. (2009). *Handbook of Nonlinear Filtering*, Oxford University Press, New York.

Du, Y., Budman, H., Duever, T.A. (2017). Comparison of stochastic fault detection and classification algorithms for nonlinear chemical processes. *Computers & Chemical Engineering*, 106, 57–70.

El Amri, M.R., Helbert, C., Lepreux, O., Zuniga, M.M., Prieur, C., Sinoquet, D. (2020). Data-driven stochastic inversion via functional quantization. *Statistics and Computing*, 30(3), 525–541.

Garud, S.S., Karimi, I.A., Kraft, M. (2017). Smart sampling algorithm for surrogate model development. *Computers & Chemical Engineering*, 96, 103–114.

Gertler, J. (1991). Analytical redundancy methods in fault detection and isolation-survey and synthesis. *IFAC Proceedings Volumes*, 24(6), 9–21.

Kalman, R.E. (1960). A new approach to linear filtering and prediction problems. *J. Basic Eng.*, 82, 35–45.

Khan, S. and Yairi, T. (2018). A review on the application of deep learning in system health management. *Mechanical Systems and Signal Processing*, 107, 241–265.

Koller, D., Friedman, N., Bach, F. (2009). *Probabilistic Graphical Models: Principles and Techniques*. MIT Press. Cambridge, MA.

Lo, N.G., Flaus, J.-M., Adrot, O. (2019). Review of machine learning approaches in fault diagnosis applied to IOT systems. *International Conference on Control, Automation and Diagnosis (ICCAD)*, Grenoble, France.

Mattis, S.A. and Wohlmuth, B. (2018). Goal-oriented adaptive surrogate construction for stochastic inversion. *Computer Methods in Applied Mechanics and Engineering*, 339, 36–60.

McBride, K. and Sundmacher, K. (2019). Overview of surrogate modeling in chemical process engineering. *Chemie Ingenieur Technik*, 91(3), 228–239.

Nor, N.M., Hassan, C.R.C., Hussain, M.A. (2020). A review of data-driven fault detection and diagnosis methods: Applications in chemical process systems. *Reviews in Chemical Engineering*, 36(4), 513–553.

Patton, R.J. and Chen, J. (1997). Observer-based fault detection and isolation: Robustness and applications. *Control Engineering Practice*, 5(5), 671–682.

Provan, G. (2017). Co-design of embeddable diagnostics using reduced-order models. *IFAC-PapersOnLine*, 50(1), 12222–12229.

Provan, G. (2019). Model-based diagnosis with probabilistic models. In *Fault Diagnosis of Dynamic Systems*, Escobet, T., Bregon, A., Pulido, B., Puig, V. (eds). Springer, Cham.

Rezaeianjouybari, B. and Shang, Y. (2020). Deep learning for prognostics and health management: State of the art, challenges, and opportunities. *Measurement*, 163, 107929.

Sharifian, S., Sotudeh-Gharebagh, R., Zarghami, R., Tanguy, P., Mostoufi, N. (2019). Uncertainty in chemical process systems engineering: A critical review. *Reviews in Chemical Engineering* [Online]. Available at: https://doi.org/10.1515/revce-2018-0067.

Smyth, P. (1994). Hidden Markov models for fault detection in dynamic systems. *Pattern Recognition*, 27(1), 149–164.

Staroswiecki, M. and Declerck, P. (1990). Analytical redundancy in non linear interconnected systems by means of structural analysis. *IFAC Proceedings Volumes*, 22, 51–55.

Stuart, A.M. (2010). Inverse problems: A Bayesian perspective. *Acta Numerica*, 19, 451–559.

Venkatasubramanian, V., Rengaswamy, R., Yin, K., Kavuri, S.N. (2003). A review of process fault detection and diagnosis: Part I: Quantitative model-based methods. *Computers & Chemical Engineering*, 27(3), 293–311.

Vogel, C.R. (2002). *Computational Methods for Inverse Problems*. SIAM, Philadelphia.

Xiong, Q. and Jutan, A. (2002). Grey-box modelling and control of chemical processes. *Chemical Engineering Science*, 57(6), 1027–1039.

Yan, L. and Zhou, T. (2019). An adaptive surrogate modeling based on deep neural networks for large-scale Bayesian inverse problems [Online]. Available at: arXiv preprint arXiv:1911.08926.

Yang, Y. and Perdikaris, P. (2019). Conditional deep surrogate models for stochastic, high-dimensional, and multi-fidelity systems. *Computational Mechanics*, 64, 417–434.

Zhang, S., Zhang, S., Wang, B., Habetler, T.G. (2019). Machine learning and deep learning algorithms for bearing fault diagnostics – A comprehensive review [Online]. Available at: arXiv preprint arXiv:1901.08247.

Zhao, R., Yan, R., Chen, Z., Mao, K., Wang, P., Gao, R.X. (2019). Deep learning and its applications to machine health monitoring. *Mechanical Systems and Signal Processing*, 115, 213–237.

5

Data-Driven Methods for Fault Diagnosis

Silvio SIMANI
Department of Engineering, University of Ferrara, Emilia-Romagna, Italy

5.1. Introduction

The problem of identifying an unknown system, given samples of its behavior, is well-known (Söderström and Stoica 1987; Kalman 1990; Ljung 1999) to be ill-posed (Hadamard 1964), as its solution is not unique and does not depend continuously on the given data.

When *a priori* knowledge on the characteristics of the unknown system is available, the identification procedure can be enhanced. This knowledge may act as a set of constraints shaping the space of possible models, so that identification problems in this new space become more tractable. As an example, the regularity of the unknown system can be translated into smoothness constraints of some kind, transforming the identification problem into a minimization problem (Tikhonov and Arsenin 1977; Morozov 1984).

This point of view can be successfully applied to estimate algebraic and dynamic affine systems from noisy samples, by assuming certain good properties of the noise and sampling process (Kalman 1982b; Beghelli and Soverini 1992).

The identification methods described in this chapter, and exploited for fault diagnosis, start from the results based on an algebraic case with the purpose of showing the possibility of extending the Frisch scheme (Frisch 1934) to dynamic systems, determining the whole family of models compatible with noisy sequences.

Diagnosis and Fault-tolerant Control 1,

Systems theory borrows some tools and viewpoints from existing and well-established fields. This happens often in new disciplines. Thus, the identification of static and dynamical systems, that is, the determination of models from noisy data, has relied heavily on techniques developed by statisticians (e.g. Frisch), who have traditionally considered it mandatory to associate a unique model to every available set of data, whether they are contaminated by noise or not.

As an example, Kalman (1982a,1982b, 1984) reconsidered this problem, pointing out how the association of a single model to uncertain data is often based on the introduction of additional information, unrelated to the data, that is, of prejudices. While the introduction of such prejudices can be convenient in some practical cases, it is, of course, very important to evaluate the family of solutions that can be found, without introducing prejudices or, at least, by only introducing mild ones.

Such an analysis has been performed by Kalman in the previously mentioned works, with reference to the Frisch scheme and to noisy data generated by linear algebraic systems. In this context, Kalman describes the class of *Errors-In-Variables* (EIV) models (see section 5.2) compatible with the given noisy sequences; his solution is in accordance with the uncertainty principle, stating that no uncertain data can lead, in general, to a unique model.

The research described in this chapter, which will be exploited for fault diagnosis, started from algebraic results with the purpose of investigating the possibility of extending the Frisch scheme to dynamical systems, determining the whole family of models compatible with noisy sequences. The results obtained differ from the expectations of the authors in that, as it is proved in the following sections, a single model is, in general, compatible with the data. This result is not in contrast with Kalman's result; in fact, in the dynamic case, the additional information necessary to obtain a single model is carried by the correlations established among the samples, by the dynamic nature of the generating process (Fantuzzi *et al.* 2002).

A frequency approach for EIV models (Kalman 1982b, 1990) and its application to the dynamic Frisch scheme identification is still in development (Beghelli *et al.* 1994a,b, 1997).

This chapter also addresses the problem of the identification of both linear and nonlinear dynamic systems for fault diagnosis. In the case of nonlinear dynamic systems, the identification will be performed by exploiting parametric nonlinear models, such as affine, piecewise-affine and fuzzy models (Fantuzzi *et al.* 2002).

Model estimation from noisy data, that is, model parameters, order and structure, is then achieved by using an extension of the so-called Frisch scheme procedure (Fantuzzi *et al.* 2002).

5.2. Models for linear system fault diagnosis

In the following, it is assumed that the monitored system can be described when working in some steady-state operating condition using a linear, discrete-time, time-invariant, dynamic model of the type

$$\begin{cases} x(t+1) = A\,x(t) + B\,u^*(t) \\ y^*(t) \quad\;\; = C\,x(t) \;\; t = 1,\, 2,\, \cdots \end{cases} \qquad [5.1]$$

where $x(t) \in \Re^n$ is the state vector, $y^*(t) \in \Re^m$ is the process output vector and $u^* \in \Re^r$ is the control input vector. A, B, C are constant matrices of appropriate dimensions, obtained by means of modeling techniques or identification procedures (recalled in this chapter for fault diagnosis applications).

The input and output variables are measured through sensors for identification purposes. Generally, sensor measurements are affected by additive noise, that can be modeled as:

$$\begin{cases} u(t) = u^*(t) + \tilde{u}(t), \\ y(t) = \; y^*(t) + \tilde{y}(t) \end{cases} \qquad [5.2]$$

where the variables $\tilde{u}(t)$ and $\tilde{y}(t)$ represent noises that affect sensor output, which are generally described as white, zero-mean, uncorrelated Gaussian noises.

It is assumed that $u(t)$ and $y(t)$ are the only available measurements from the real process. It is worth noting that representations of types, such as equations [5.1] and [5.2], are known as EIV models.

Two conditions were considered regarding the identification of parametric models characterizing system behavior, namely, high and low signal-to-noise ratios.

In the case of high signal-to-noise ratios ($\tilde{u}(t) \cong 0$ and $\tilde{y}(t) \cong 0$), Equation Error (EE) identification can be exploited and, in particular, different EE models can be extracted from the data. A specific discrete-time, time-invariant, linear dynamic model, for example, ARX or ARMAX (Auto Regressive eXogenous or Auto Regressive Moving Average eXogenous) (Söderström and Stoica 1987; Ljung 1999) can only be selected inside an assumed family of models.

On the other hand, if the signal-to-noise ratios on the input and output of the process are low, the Frisch scheme (Frisch 1934) can be applied to perform dynamic system identification. Such a scheme facilitates the determination of the linear discrete system that generates the sequences, as well as the variances of the noises $\tilde{u}(t)$ and $\tilde{y}(t)$ affecting the data (Beghelli *et al.* 1990). In the Frisch scheme, these signals are assumed to be zero-mean white noises, mutually uncorrelated and uncorrelated with every component of $u^*(t)$ and $y^*(t)$.

The input to output discrete-time model behavior can be mathematically described by a set of ARX multi-input single-output (MISO) models of the type:

$$y_i^*(t) = \sum_{j=1}^{n} \alpha_{i,j} y_i^*(t-j) + \sum_{j=1}^{r} \sum_{k=1}^{n} \beta_{i,j,k} u_j^*(t-k) + \varepsilon_i(t), \quad i = 1, \ldots, m \qquad [5.3]$$

where m is the number of output variables. The order n and the parameters $\alpha_{i,j}$ and $\beta_{i,j,k}$, with $i = 1, \cdots, m$, of the model have to be determined by the identification approach. The term $\varepsilon_i(t)$ takes into account the modeling error, which is due to process noise, parameter variations, etc.

In either case, the subsequent steps consist of transforming the input-output discrete-time time-invariant linear models from equation [5.3] into state-space representations.

The state-space systems obtained by the EE models are useful to design dynamic observers (Luenberger 1971), while the ones coming from the Frisch scheme can be used to build Kalman filters (Jazwinski 1970).

As presented by Söderström and Stoica (1987), the input-output EE model from equation [5.3] has a state-space realization as follows:

$$\begin{cases} x_i(t+1) = A_i\, x_i(t) + B_i\, u^*(t) + B_{\omega_i}\, \varepsilon_i & i = 1, \cdots, m \\ y_i^*(t) \quad\;\; = C_i\, x_i(t) + D_{\omega_i}\, \varepsilon_i, & t = 1, 2, \cdots \end{cases} \qquad [5.4]$$

where the matrices A_i $(n \times n)$, B_i $(n \times r)$, B_{ω_i} $(n \times 1)$, C_i $(1 \times n)$ and D_{ω_i} depend on system order and on $\alpha_{i,j}$ and $\beta_{i,j,k}$ parameters.

The matrices $(A_i, B_i, C_i, B_{\omega_i}, D_{\omega_i})$ of a state-space representation, in canonical form of the n-th order system given by equation [5.4], can be defined as follows:

$$A_i = \begin{bmatrix} 0 & 1 & 0 & \cdots & 0 \\ 0 & 0 & 1 & \cdots & 0 \\ \vdots & \vdots & \vdots & \ddots & \vdots \\ \alpha_{i,1} & \alpha_{i,2} & \alpha_{i,3} & \cdots & \alpha_{i,n} \end{bmatrix},$$

$$C_i = \begin{bmatrix} 1\, 0 \cdots 0 \end{bmatrix}, \qquad [5.5]$$

and:

$$\begin{bmatrix} B_i & B_{\omega_i} \\ 0 & D_{\omega_i} \end{bmatrix} = S_i^{-1} \begin{bmatrix} \beta_{i,1,1} & \cdots & \beta_{i,r,1} & 0 \\ \vdots & \vdots & \vdots & \vdots \\ \beta_{i,1,n} & \cdots & \beta_{i,r,n} & 0 \\ 0 & \cdots & 0 & 1 \end{bmatrix} \qquad [5.6]$$

where the S_i $((n+1) \times (n+1))$ matrix is defined as:

$$S_i = \begin{bmatrix} -\alpha_{i,1} & -\alpha_{i,2} & \cdots & -\alpha_{i,n} & 1 \\ -\alpha_{i,2} & -\alpha_{i,3} & \cdots & 1 & 0 \\ \vdots & \vdots & \vdots & \ddots & \vdots \\ -\alpha_{i,n} & 1 & 0 & 0 & 0 \\ 1 & 0 & 0 & 0 & 0 \end{bmatrix} \qquad [5.7]$$

Section 5.3 discusses procedures for obtaining the needed parameters, based on the model structures outlined above.

5.3. Parameter estimation methods for fault diagnosis

Many classes of models can be considered from an identification standpoint that can be used for fault diagnosis; the most relevant descriptions considered in this work are the discrete-time, parametric, deterministic and stochastic, input-output and MISO dynamic models.

Real systems are always affected by disturbances (noise entering into the system and/or affecting the measures, unknown inputs, etc.). These disturbances or their global effect can be described by means of noises acting on the input, state and output of the model, which can be called, in this case, stochastic. Often, the global effect of disturbances can be modeled as the output of a filter driven by white noise, and is added to the output of the deterministic part of the model, which is thus decomposed into a deterministic and a stochastic part. Depending on the applications, it can be sufficient to identify the deterministic part of the model (high signal-to-noise ratio case, section 5.3.1), or it can be necessary to identify both parts (low signal-to-noise ratio, section 5.3.2).

5.3.1. *Data–driven method in ideal conditions*

This section considers a SISO (single-input single-output, $m = 1$ and $r = 1$) ARX model with order n (see equation [5.3]), and the input-output sequences $\{u(t), y(t)\}$

observed in the time interval $[1, L]$. If model [5.3] is used to compute the predicted output values $y^*(t)$ in the $N = L - n$ times, for a given set of parameters:

$$\theta = \begin{bmatrix} \alpha_n \cdots \alpha_1 \; \beta_n \cdots \beta_1 \end{bmatrix} \qquad [5.8]$$

the mean–square prediction error $J(\theta)$ is given by:

$$J(\theta) = \frac{1}{N}\sum_{t=n+1}^{L} \left(y^*(t) - y(t)\right)^2 . \qquad [5.9]$$

By introducing the following Hankel matrices H_u and H_y:

$$H_n(u) = \begin{bmatrix} u(1) & \cdots & u(n) \\ \vdots & \ddots & \vdots \\ u(L-n) & \cdots & u(L-1) \end{bmatrix} \text{ and}$$

$$H_n(y) = \begin{bmatrix} y(1) & \cdots & y(n) \\ \vdots & \ddots & \vdots \\ y(L-n) & \cdots & y(L-1) \end{bmatrix} \qquad [5.10]$$

it follows that:

$$\begin{bmatrix} y^*(n+1) \\ \vdots \\ y^*(L) \end{bmatrix} = \left[H_n(y) H_n(u)\right] \theta = H_n \theta \qquad [5.11]$$

It can be proved that the parameter vector minimizing the cost function [5.9] is given by:

$$\hat{\theta} = H_n^+ \begin{bmatrix} y(n+1) \\ \vdots \\ y(L) \end{bmatrix} = H_n^+ y_n^o \qquad [5.12]$$

where H_n^+ denotes the pseudo-inverse of the H_n matrix. The algorithm gives an estimate $\hat{\theta}$ of θ, which converges asymptotically to the real parameter of the process that has generated the data.

To estimate the order n of the ARX process, an integer $k > 0$ and the $(N \times 2k+1)$ matrix of input-output samples given by:

$$H_k^* = [H_k(y) H_k(u) y_k^o] \quad [5.13]$$

are considered.

If $\varepsilon_i(t) = 0$ in equation [5.3], the following properties hold:

$$\begin{array}{lll} \text{rank}\,(H_k^*) = 2k+1 & \text{for } 2k+1 < n \\ \text{rank}\,(H_k^*) = 2k & \text{for } 2k \geq n \end{array} \quad [5.14]$$

It could be possible to consider the increasing sequence of matrices:

$$S_1\ S_2\ \cdots\ S_n\ \cdots \quad [5.15]$$

where $S_k = H_k^{*T} H_k^*$ to evaluate their singularity.

The first singular matrix S_k would define the correct order for the model $(k = n)$. Unfortunately, the presence of $\varepsilon_i(t) \neq 0$ in equation [5.3] leads to the non-singularity of every matrix in equation [5.15].

It can be proven that if N is large enough, an estimate of the standard deviation σ_ε of the process $\varepsilon(t)$ in equation [5.3], is given by Ljung (1999):

$$\sigma_\varepsilon = \sqrt{\frac{\det(S_n)}{N \det(H_n^T H_n)}} \quad [5.16]$$

In the following, the quantity σ_{ε_k} is defined:

$$\sigma_{\varepsilon_k} = \sqrt{\frac{\det(S_k)}{N \det(H_k^T H_k)}} \quad [5.17]$$

It can be shown that $\sigma_{\varepsilon_h} > \sigma_{\varepsilon_k}$ for $h < k$, $\sigma_{\varepsilon_k} > \sigma_\varepsilon$ for $k < n$ and $\sigma_{\varepsilon_k} \cong \sigma_\varepsilon$ for $k \geq n$. In other words, if N is large enough, a sequence of decreasing values of σ_{ε_k}, followed by a stabilization once the correct order is reached, can be noted. The criterion can be used to evaluate a suitable order or, at least, an interval of admissible orders for the model before computing its parameters. Moreover, $\sigma_{\varepsilon_k}^2 = J_k(\theta)$ for an ARX model with order k and parameters θ given by equation [5.12].

If the value in equation [5.17] is expressed as a percentage of the standard deviation of the measured output, the well-established Predicted Per Cent Reconstruction Error criterion (PPCRE) is obtained (Guidorzi *et al.* 1982).

The PPCRE(k) gives the prediction error of an ARX model of order k without requiring any computation of its parameters and predictions.

The application of the PPCRE criterion consists of computing an increasing sequence of PPCRE(k) (or $J_k(\theta)$) and selecting the minimal order that, once increased, does not lead to a significantly better performance.

See Figures 5.1 and 5.2 for an example of the PPCRE(k) and $J_k(\theta)$ increasing sequences with ARX model order k for a simulated second-order model ($n = 2$).

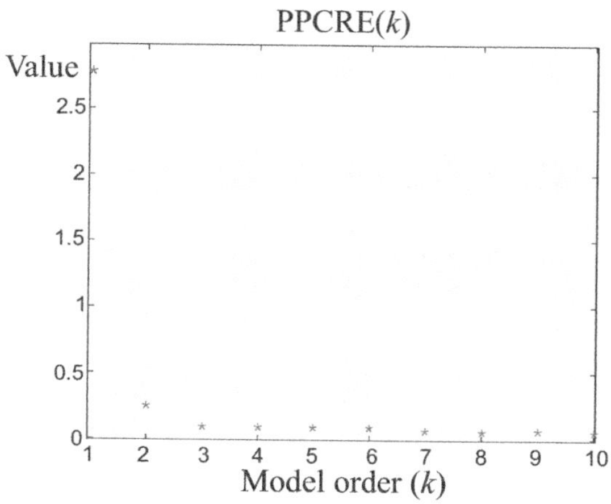

Figure 5.1. *PPCRE criterion for different ARX model orders* k

The change in the slope is clear for both of the curves represented in Figures 5.1 and 5.2 corresponding to the order $n = 2$. The relation from equation [5.17] can also be used in the application of the well-established FPE, AIC and MDL order estimation criteria (Söderström and Stoica 1987).

5.3.2. *Data-driven methods in real scenarios*

In this section, the identification of linear models from noisy data is presented. In the interest of clarity, we consider the algebraic and dynamic cases separately.

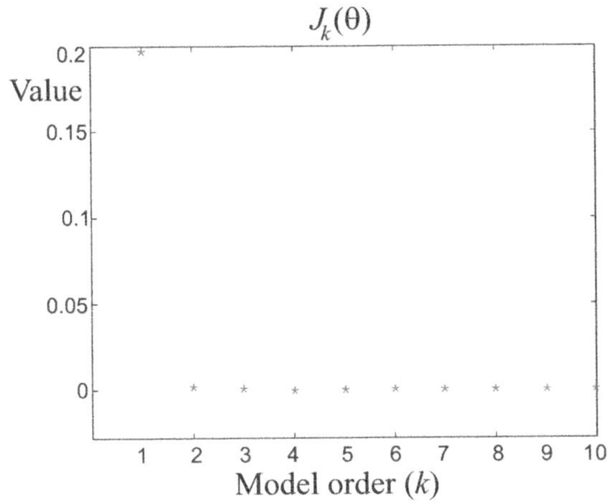

Figure 5.2. *Mean square error criterion for different ARX model orders k*

5.3.3. *Algebraic Frisch scheme*

We consider the finite sequence of n variables x_1, x_2, $\cdots$, x_n observed at N different times with $N > n$. If linear relations exist among these variables, they are described by models of the type:

$$a_1 x_1 + a_2 x_2 + \cdots + a_n x_n = 0 \quad [5.18]$$

If X is the $(N \times n)$ matrix storing the variable samples, model [5.18] is described by the columns of a matrix A such that:

$$X\,A = 0 \quad [5.19]$$

or, equivalently:

$$X^T\,X\,A = 0 \quad [5.20]$$

where $\Sigma = X^T X$ is the sample covariance matrix, under a zero-mean assumption for all variables.

When the data are corrupted by noise, the rank$[\Sigma] = n$, so that no relation can be obtained unless the data are modified. This explains the reason for using the EIV models. In the Frisch scheme, the following assumptions are therefore added:

1) All variables are treated symmetrically and each variable is affected by an unknown amount of additive noise.

2) Each noise component is independent of every other noise component and of every variable. Under these conditions, each variable x_i, with $i = 1, \cdots, n$, is defined as

$$x_i = x_i^* + \tilde{x}_i \quad [5.21]$$

where the unknown terms x_i^* are the true value of the ith variable, and $\tilde{x}_i$ are the additive noises on this variable.

The problem of determining the true data from the available noisy sequences can thus be formulated as follows. Given an $(n \times n)$ symmetric positive–definite covariance matrix Σ, find all diagonal matrices $\tilde{\Sigma}$ with non-negative elements, such that $\Sigma^* = \Sigma - \tilde{\Sigma} \geq 0$. In this context, the solution of the problem is not unique.

The rank of Σ^* may change by varying $\tilde{\Sigma}$, which models the noise and, consequently, the same set of data may be linked by different numbers of linear relations.

Even if it is not the case, the problem has infinite solutions; the rank of Σ^*, for instance, is always equal to $(n - 1)$ if and only if Σ^{-1} can be reduced to a matrix with strictly positive entries by the transformation $L\Sigma^{-1}L$ with $L = \text{diag}[\pm 1]$.

The solution set is the convex simplex, the n vertices of which are the least-squares solutions, which can be found by assuming that one variable is noisy and all of the others are noise-free (Beghelli *et al.* 1990).

Before solving the previous problem, the following result can be considered (Beghelli *et al.* 1990).

Given the $(n \times n)$ symmetric positive-definite covariance matrix Σ, the maximal variance of the additive noise on the ith variable, when all others are noise-free, is computed by:

$$\tilde{\sigma}_i = \frac{\det[\Sigma]}{\det[\Sigma_i]} \quad [5.22]$$

where Σ_i is obtained from Σ by deleting its ith row and column.

This achievement means that every allowable noise covariance matrix $\tilde{\Sigma}$ (i.e., such that $\Sigma^* = \Sigma - \tilde{\Sigma} \geq 0$) defines a point $(\sigma_1, \cdots, \sigma_n)$ belonging to the first orthant of the noise space $\Re^n$, which is mapped into one and only one point $(a_1, \cdots, a_n)$ of the solution space $\Re^n$. Moreover, the following solution can be also considered (Beghelli *et al.* 1990).

The set defined by all points $(\sigma_1, \cdots, \sigma_n)$, defined by the matrix set $\tilde{\Sigma}$ is a convex hypersurface belonging to the first orthant of the noise space, the section of which, with a plane parallel to a coordinate one, is a hyperbola segment. Note that if noise values, corresponding to a rank of Σ^* lower than $n-1$, exist, they belong to the hypersurface defined by the previous result. In these conditions, in the parameter space, the solution set might be a collection of convex polyhedral sets lying in the orthants.

The hypersurface defined by the previous statements partitions the first orthant of the noise space into two regions. The points over the hypersurface correspond to non-definite matrices Σ^*, those under the hypersurface to positive-definite matrices.

5.3.4. *Dynamic Frisch scheme*

Consider a finite sequence of the variables $u_1(\cdot)$, $\cdots$, $u_r(\cdot)$, $y(\cdot)$ observed with a constant sampling interval. If dynamic linear relations exist among these variables, they can be described by models of the type:

$$y^*(t+n) = \sum_{i=0}^{n-1} \alpha_i y^*(t+i) + \sum_{i=0}^{n-1} \sum_{j=1}^{r} \beta_{ij} u_j^*(t+i) \qquad [5.23]$$

which represent linear MISO discrete-time systems whose order is n and whose parameters are α_i and β_{ij}.

The following problem is considered first. Given a noiseless input-output sequence $u_1^*(\cdot)$, $\cdots$, $u_r^*(\cdot)$, $y^*(\cdot)$ generated by a system in the form given by equation [5.23], determine the order n and the parameters α_i, β_{ij} of the system.

The following vectors and matrices are thus defined as:

$$u_j^{*N}(t+k) = \left[u_j^*(t+k) \ldots u_j^*(t+k+N-1) \right]^T \qquad [5.24]$$

$$y^{*N}(t+k) = \left[y^*(t+k) \ldots y^*(t+k+N-1) \right]^T \qquad [5.25]$$

$$X_k(u_j^*) = \left[u_j^{*N}(t) \ldots u_j^{*N}(t+k-2) \right] \qquad [5.26]$$

$$X_k(y^*) = \left[y^{*N}(t) \ldots y^{*N}(t+k-1) \right] \qquad [5.27]$$

$$\Sigma_k^*(u_j^* u_j^*) = X_k^T(u_j^*) X_k(u_j^*) \qquad [5.28]$$

$$\Sigma_k^*(y^* y^*) = X_k^T(y^*) X_k(y^*) \qquad [5.29]$$

$$\Sigma_k^*(y^* u_j^*) = X_k^T(y^*) X_k(u_j^*) = \Sigma_k^{*T}(u_j^* y^*) \qquad [5.30]$$

where N is assumed to be large enough to solve the problem considered.

Let us now partition the matrix Σ_k^* as follows:

$$\Sigma_k^* = \begin{bmatrix} \Sigma_k^*(y^*y^*) & \Sigma_k^*(y^*u_1^*) & \dots & \Sigma_k^*(y^*u_r^*) \\ \Sigma_k^*(u_1^*y^*) & \Sigma_k^*(u_1^*u_1^*) & \dots & \Sigma_k^*(u_r^*y^*) \\ \vdots & \vdots & \ddots & \vdots \\ \Sigma_k^*(u_r^*y^*) & \Sigma_k^*(u_r^*u_1^*) & \dots & \Sigma_k^*(u_r^*u_r^*). \end{bmatrix}. \qquad [5.31]$$

To solve the so-called realization problem, it is possible to consider the sequence of increasing-dimension matrices:

$$\Sigma_2^*, \ \Sigma_3^*, \ \dots \ \Sigma_k^*, \ \dots \qquad [5.32]$$

testing their singularity. As soon as a singular matrix Σ_k^* is found, then $n = k - 1$ and the parameters $\alpha_0, \dots, \alpha_{n-1}, \beta_{0j}, \dots, \beta_{(n-1)j}$ $(j = 1, \dots, r)$ describe the dependence relationship of the $(n+1)$-th vector of Σ_{n+1}^* on the remaining ones.

It has been assumed that N is large enough to avoid unwanted linear dependence relationships, due to limitations in the dimension of the involved vector spaces; this means $N \geq (r+1)n+1$. If a lower number of samples is available then only a partial realization problem can be solved.

In the noisy case, the following identification problem can be proposed. Given a noisy input-output sequence $u_1(\cdot)$, ..., $u_r(\cdot)$, $y(\cdot)$, uniquely determine, if possible, the order n and the parameters α_i, β_{ij} of a model, as shown in equation [5.23] of the system, which has generated the noiseless sequences $u_1^*(\cdot)$, ..., $u_r^*(\cdot)$, $y^*(\cdot)$.

Note that in the presence of noise, the procedure described for the solution of the noiseless identification problem would obviously be useless since the matrices Σ_k^* would always be non-singular.

In the Frisch scheme, it is normally assumed that:

$$\begin{cases} u_j(t) = u_j^*(t) + \tilde{u}_j(t), & j = 1, \dots, r \\ y(t) \ = y^*(t) + \tilde{y}(t) \end{cases} \qquad [5.33]$$

where every noise term $\tilde{u}_j(t)$, $\tilde{y}(t)$ is independent of every other term and only $u_j(t)$ and $y(t)$ are known. Without a loss of generality, all of the variables may be assumed to be zero-mean.

Consequently, the generic positive-definite matrix Σ_k associated with the input-output noise-corrupted sequences may always be expressed as the sum of two terms $\Sigma_k = \Sigma_k^* + \tilde{\Sigma}_k$ where:

$$\tilde{\Sigma}_k = \text{diag}[\tilde{\sigma}_y I_k, \tilde{\sigma}_{u_1} I_{k-1}, \dots, \tilde{\sigma}_{u_r} I_{k-1}] \geq 0 \qquad [5.34]$$

since no correlation has been assumed among the noise samples at different times. This condition is verified for additive white noise, with variance $\tilde{\sigma}_y$ and $\tilde{\sigma}_{u_j}$ on the input-output sequences.

In the noisy scenario, the following problem has to be investigated. Given a sequence of increasing–dimension $\Big(\big((r+1)k-r\big) \times \big((r+1)k-r\big)\Big)$ symmetric positive-definite covariance matrices with:

$$\Sigma_2, \ \Sigma_3, \ \ldots \ \Sigma_k, \ \ldots \qquad [5.35]$$

find, for each k, all diagonal non-negative definite matrices:

$$\tilde{\Sigma}_k = \text{diag}[\tilde{\sigma}_y I_k, \tilde{\sigma}_{u_1} I_{k-1}, \ldots, \tilde{\sigma}_{u_r} I_{k-1}] \qquad [5.36]$$

such that:

$$\Sigma_k^* = \Sigma_k - \text{diag}[\tilde{\sigma}_y I_k, \tilde{\sigma}_{u_1} I_{k-1}, \ldots, \tilde{\sigma}_{u_r} I_{k-1}] \geq 0 \qquad [5.37]$$

Note that, unlike the algebraic case, for each k, the noise space is always $\Re_+^{(r+1)}$, while the parameter space is defined on $\Re^{(r+1)k-r}$.

It can be seen that for each k, the solution set of the relation from equation [5.37] describes, in the first orthant of the $(\tilde{\sigma}_y, \tilde{\sigma}_{u_1}, \ldots, \tilde{\sigma}_{u_r})$ hyperplane, a hypersurface whose concavity faces the origin (Beghelli *et al.* 1990).

In the noise space, the $(r+1)$ solutions $(\tilde{\sigma}_y, 0, \tilde{\sigma}_{u_2}, \ldots, \tilde{\sigma}_{u_r})$, $(\tilde{\sigma}_y, \tilde{\sigma}_{u_1}, 0, \ldots, \tilde{\sigma}_{u_r})$, $\cdots$, $(\tilde{\sigma}_y, \tilde{\sigma}_{u_1}, \ldots, \tilde{\sigma}_{u_{r-1}}, 0)$ correspond to the limit case of noise affecting only the output or the input sequences. This case can be considered as the natural extension to the dynamic case of the computation of least-square solutions.

Previous results hold for every value of k. Since the determination of the system order requires the increasing values of k to be tested, it is relevant to analyze the behavior of the associated curves when k varies.

This corresponds to a comparison of the admissible solution sets for different model orders. In this context, the following result can be proved (Beghelli *et al.* 1990). The solution sets of the condition from equation [5.37] for different values of k are non-crossing curves.

It is also important to observe that, since it is assumed that a system of the form given by equation [5.23] has generated the noiseless data, for $k > n$ all of the hyper-surfaces of the type described by equation [5.37] have at least one common point, that is, point $(\tilde{\sigma}_{y^*}, \tilde{\sigma}_{u_1^*}, \ldots, \tilde{\sigma}_{u_r^*})$, corresponding to the true variances $\tilde{\sigma}_{y^*}$ and $\tilde{\sigma}_{u_j^*}$ of the noise affecting the output and inputs of the system.

The search for a solution for the identification problem can thus start by determining this point in the noise space. The following considerations can now be stated.

With reference to the diagonal non-negative definite matrices:

$$\tilde{\Sigma}_k = \text{diag}[\tilde{\sigma}_{y^*} I_k, \tilde{\sigma}_{u_1^*} I_{k-1}, \ldots, \tilde{\sigma}_{u_r^*} I_{k-1}] \qquad [5.38]$$

the following properties hold:

– if $k \leq n$, the matrices Σ_k^* are positive-definite;

– if $k > n$, the dimension of the null space of Σ_k^* and, consequently, the multiplicity of its least eigenvalue, is equal to $(k - n)$;

– for $k = (n + 1)$, the matrix Σ_k^* is characterized by a linear dependence relation among its $(r+1)k-r$ vectors, and the coefficients which link the k-th vector of Σ_k^* to the remaining ones are the parameters α_i, β_{ij}, with $i = 0, \ldots, n-1$ and $j = 1, \ldots, r$, of the system given by equation [5.23], which has generated the noiseless sequences;

– for $k > (n + 1)$, all of the linear dependence relations among the vectors of the matrix Σ_k^* are characterized by the same $(r + 1)n$ coefficients α_i, β_{ij}.

As an example, Figure 5.3 shows the above properties for a second order ($n = 2$) SISO dynamic system.

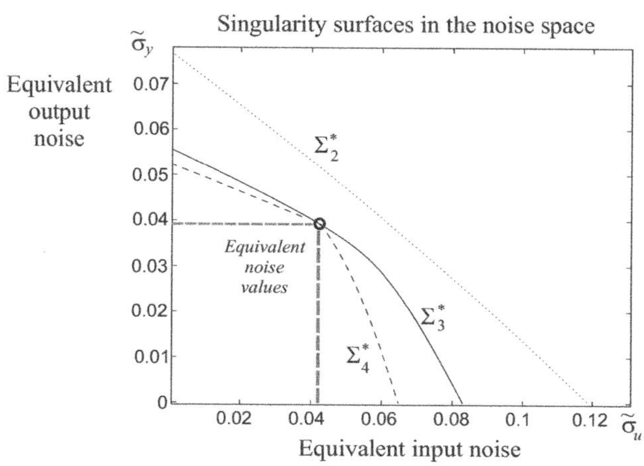

Figure 5.3. *Noise space and singularity surfaces*

The point marked by a circle corresponds to the input-output noise variances $\tilde{\sigma}_{y^*}$ and $\tilde{\sigma}_{u^*}$ affecting the measurements.

It is worth observing how this approach cannot be applied immediately to the identification of real processes, since the hypotheses on the linearity, finite dimensionality and time independence of the system and on the additivity and whiteness of the noise are not usually verified. Therefore, the hyper-surfaces defined by equation [5.37] have no common point for $k > n$. Finally, note that the definition of a suitable criterion of model selection in such cases was suggested in Beghelli *et al.* (1994a).

5.3.5. *MIMO case Frisch scheme*

A multivariable system can be represented by means of canonical models of the form:

$$y_i^*(t + \nu_i) = \sum_{j=1}^{m} \sum_{k=0}^{\nu_{ij}-1} \alpha_{ijk} y^*(t + k) + \sum_{j=1}^{r} \sum_{k=0}^{\nu_i - 1} \beta_{ijk} u_j^*(t + k) \qquad [5.39]$$

where $i = 1, \cdots, m$. r and m are the number of inputs and outputs of the system, respectively (Guidorzi 1981). The indices $\nu_{ij} - 1$ satisfy the following relations:

$$\begin{cases} \nu_{ij} = \nu_i & \text{for } i = j \\ \nu_{ij} = \min(\nu_i + 1, \nu_j) & \text{for } i > j \\ \nu_{ij} = \min(\nu_i, \nu_j) & \text{for } i < j \end{cases} \qquad [5.40]$$

The model from equation [5.39] decomposes the system into m interconnected subsystems, the orders of which are given by the integers ν_i. Such integers completely define the system structure and are coincident with the Kronecker observability invariants of any realization of the system.

Given the sequences $y_i(t)$ $(i = 1, \cdots, m)$ and $u_j(t)$ $(j = 1, \cdots, r)$ generated by the system of the type given by equation [5.39], the identification problem consists of uniquely determining both the *structure*, that is, the set of integers $\nu_1, \cdots, \nu_m$, as well as the characteristic *parameters* α_{ijk} and β_{ijk} of the model from equation [5.39].

The solution of the identification problem is described in Guidorzi (1981), with reference to canonical models, but can easily be generalized to multistructural (overlapping) models. Advantages associated with the use of these identified models with reference to FDI concern the minimal parameterization (Delmaire *et al.* 1999), reduced storage, computing time and high efficiency of the related algorithms.

The techniques and properties for MISO system identification, presented in section 5.3.4, can be generalized for the MIMO case. Because of these properties, it is possible to conclude that if, starting from a certain structure, the hypersurfaces associated with the increasing dimension of covariance matrices only have a common point in the noise space, then this point represents the variances of the noise affecting the input-output sequences. If the noise variances can thus be uniquely determined, the identification problem can be reduced to that of a realization and a canonical model of the system can be obtained.

From a computational point of view, it can be noted that the search for the noise variances is made in an $(m + r)$-dimensional space and may, therefore, be time expensive.

The results given in section 5.3.4 and extended to the MIMO case do not exclude the possibility that the previously considered hypersurfaces may be coincident and consequently that a non-unique solution exists.

In conclusion, it can be stated that no conceptual differences exist between the application of the Frisch scheme to the identification of SISO and MIMO dynamic systems.

5.4. Nonlinear dynamic system identification

In this section the problem of finding a suitable nonlinear model for real processes is presented and used for the purpose of fault diagnosis. With the aim of providing a general treatment for the problem, our attention is focused on nonlinear "black-box" parametric models.

In fact, input-output process measurements can be successfully used to infer an analytical description of the system in the framework of a parametric structure, which presents approximation properties with respect to the complex, nonlinear, unknown analytical functions that are amenable as candidates to describe the real behavior of the observed process (Juditsky *et al.* 1995).

So, the choice of the parametric structure becomes an important and difficult step toward system identification, especially when the behavior of the target process is nonlinear, as is common by far in real world applications (Billings and Voon 1983b).

The approach chosen can be applied to nonlinear processes that operate at different regimes, as this is the usual assumption in several engineering fields, in which distinct models are associated with each admissible operating condition.

A switching function governs the transition among different models or interpolations of models. Such mathematical descriptions are referred to in current literature as *piecewise models* or *hybrid models*, as suggested by several authors (Billings and Voon 1986; Chen and Billings 1989; Chen *et al.* 1990; Friedmann 1991; Hathaway and Bezdek 1983; Johansen and Foss 1993; Skeppstedt *et al.* 1992).

In particular, in current literature great attention has been paid to the piecewise linear systems, formed by a collection of *linear* (Bemporad and Morari 1999) or *affine* (Pettit and Wellstead (1995); Sontag (1981); Banks and Kathur (1989)) *dynamic local models* and a switching function that selects the appropriate model depending on the operating point.

The following concerns a structure that merges dynamic, time-invariant and affine submodels describing the behavior of the observed process in its different operating regimes locally. This type of model was formerly introduced by Priestly (1988) as a state-dependent model and was used in a stochastic environment for time series model fitting. In particular, we discuss the interpolation properties of such models with respect to nonlinear discrete-time regression functions (Leontaritis and Billings 1985b, 1985a).

The main result shows that the hybrid models with affine submodels can approximate, with arbitrary precision, the nonlinear target function in the *Sobolev's norm* (i.e. interpolate the target function and its first order derivative).

Particular emphasis has been given to the piecewise affine model identification when data are corrupted by additive noise. This approach differs from the cited bibliography as it uses the EIV structure, with the assumption suggested by Frisch (1934) and revisited by Kalman (1982b) regarding the stochastic characterization of the noises corrupting the data. In this context, it is possible to generalize some of the basic results on the single model proposed in Beghelli *et al.* (1990), with regard to model structure and parameter identification with measurement noise rejection.

5.4.1. *Piecewise affine model*

The main idea underlying the mathematical description of nonlinear dynamic systems for fault diagnosis is based on the interpretation of single-input single-output, nonlinear, time-invariant regression models (Leontaritis and Billings 1985a, 1985b) such as:

$$y(t+n) = F\big(y(t+n-1), \cdots, y(t), u(t+n-1), \cdots, u(t)\big), \quad t = 0, 1, \ldots \qquad [5.41]$$

where $u(\cdot)$ and $y(\cdot)$ belong to the bounded input $\mathcal{U}$ and output $\mathcal{Y}$ sets, respectively, n is the finite system memory (i.e., the model order) and $F(\cdot)$ is a continuous nonlinear function defining a hypersurface from a $\mathcal{D}_n$ to $\mathcal{Y}$, with $\mathcal{D}_n$ being the Cartesian product $\mathcal{U}^n \times \mathcal{Y}^n$.

The identification of the nonlinear system can be translated to the approximation of the mathematical model given by equation [5.41], using a parametric structure that exhibits arbitrary accuracy interpolation properties.

A piecewise model defined through the composition of simple models having local validity is the natural candidate to perform this task, as it combines function interpolation properties with mathematical tractability.

In the following the proposed piecewise structure is defined, and its properties in terms of interpolation characteristics of arbitrary nonlinear functions are discussed.

5.4.2. *Hybrid model structure*

The piecewise SISO model consists of a collection of parametric submodels of the type:

$$y(t+n) = \sum_{j=0}^{n-1} \alpha_j^{(i)} y(t+j) + \sum_{j=0}^{n-1} \beta_j^{(i)} u(t+j) + b^{(i)}, \quad t = 0,\, 1,\, \ldots \qquad [5.42]$$

in which the system operating point is described by the input and output samples $y(t+n-1), \cdots, y(t)$ and $u(t+n-1), \cdots, u(t)$, that can be collected with a vector $x_n(t) = [y(t), \cdots, y(t+n-1), u(t), \cdots, u(t+n-1)]^T$. The *switching* function $\chi_i(x_n(t)), i = 1, \ldots, M$ has the form:

$$\chi_i(x_n(t)) = \begin{cases} \chi_i(x_n(t)) = 1 \text{ if } x_n(t) \in \mathcal{A}_n^{(i)} \\ \chi_i(x_n(t)) = 0 \text{ otherwise} \end{cases} \qquad [5.43]$$

where $\{R_n^{(1)}, \ldots, R_n^{(M)}\}$ is a partition of $\mathcal{D}_n$, whose structure will be characterized in the following.

Thus, the output $y(t+n)$ of the nonlinear dynamic system described by equation [5.41] can be approximated by the *piecewise affine model* $f(\cdot)$ in the form:

$$y(t+n) = f(x_n(t)) = \sum_{i=1}^{M} \chi_i(x_n(t)) [x_n(t), 1]^T a^{(i)} \qquad [5.44]$$

where the model parameters are collected in the vector $a^{(i)} = [\alpha_0^{(i)}, \ldots, \alpha_{n-1}^{(i)}, \beta_0^{(i)}, \ldots, \beta_{n-1}^{(i)}, b^{(i)}]^T$. It is worth noting that the model is affine in each $R_n^{(i)}$, with $a^{(i)}$ being the affine submodel parameters.

5.4.3. *Nonlinear system approximation*

In the previous section we proposed a piecewise affine function that aims to interpolate an arbitrary nonlinear model, conceptually enabling us to capture the behavior of any physical process. Thus, it is now very important to understand the approximation capabilities of this piecewise affine model with respect to the target function.

In particular, our interest goes beyond the simple approximation capabilities under the usual Lebesgue norm, but is also focused on the first derivative of the target function, as this avoids any jitteriness and actually assures higher approximation quality.

A metric structure is provided to the set of functions $F : \mathcal{D}_n \mapsto \mathcal{Y}$, first defining the usual Lebesgue norm of order w as:

$$\|F\|_w = \begin{cases} \left[\int_{\mathcal{D}_n} |F(x_n)|^w dx_n\right]^{\frac{1}{w}} & \text{if} \quad 1 \leq w < \infty \\ \sup_{x_n \in \mathcal{D}_n} |F(x_n)| & \text{if} \quad w = \infty \end{cases} \qquad [5.45]$$

and then a stricter Sobolev's norm, which also takes the first derivatives of the function into account

$$^1\|F\|_w = \|F\|_w + \sum_{j=1}^{2n} \left\| \frac{\partial F}{\partial x_n^j} \right\|_w \qquad [5.46]$$

with x_n^j being the jth component of vector x_n.

If $\mathcal{C}^v$ is the set of those functions $f : \mathcal{D}_n \mapsto \mathcal{Y}$ featuring continuous v-th order derivatives, the following result can be stated.

The set of input-output relationships f given by equation [5.44] is dense in $\mathcal{C}^1$ in the sense of $^1\|\cdot\|_w$ for any $w > 0$. Thus, for any $\epsilon > 0$, any $w > 0$ and any $F \in \mathcal{C}^1$ there is a piecewise affine system f such that:

$$^1\|F - f\|_w < \epsilon \qquad [5.47]$$

The proof of the previous property is based on the result on Ritz's piecewise polynomial function, which leads to the formulation of the following theorem, adapted from Strang and Fix (1973, Theorem 3.1).

Let $F : \mathcal{D}_n \mapsto \mathcal{Y}$ be given such that $F \in \mathcal{C}^1$. Also assume that the value of F is given at certain points in $\mathcal{D}_n$ called nodes and that $\mathcal{D}_n$ itself is partitioned into convex regions between these nodes. Moreover, a polynomial function f interpolating the values of F at the nodes is given in each of these regions.

If the functional form of f is such that any first-order polynomial could be exactly reproduced in each region, then for each $\epsilon > 0$ a length $\hat{d}$ exists such that, in any non-degenerating region γ in which any two points can be connected with a segment of length less than $\hat{d}$, the following relation holds:

$$\max_{x_n \in \gamma} |F(x_n) - f(x_n)| \leq \epsilon \qquad [5.48]$$

and:

$$\max_{x_n \in \gamma} \left| \frac{\partial F}{\partial x_n^j}(x_n) - \frac{\partial f}{\partial x_n^j}(x_n) \right| \leq \epsilon,$$

$$j = 1, \ldots, 2n$$

Moreover, if F has a bounded Hessian $\nabla\nabla F$, $\hat{d}$ has a simple relationship with ϵ, since two constants C_0 and C_1 exist such that:

$$\max_{x_n \in \gamma} |F(x_n) - f(x_n)| \leq C_0 \hat{d}^2 \|\nabla\nabla F\|_\infty \qquad [5.49]$$

and:

$$\max_{x_n \in \gamma} \left| \frac{\partial F}{\partial x_n^j}(x_n) - \frac{\partial f}{\partial x_n^j}(x_n) \right| \leq C_1 \hat{d} \|\nabla\nabla F\|_\infty$$

$$j = 1, \ldots, 2n,$$

where $\|\nabla\nabla F\|_\infty = \max_{x_n \in \mathcal{D}_n} \max_{j,i} \left| \frac{\partial^2 f}{\partial x_n^j x_n^i}(x_n) \right|, \quad j, i = 1, \ldots, 2n$

As $\mathcal{D}_n$ is compact, the previous result is sufficient to prove density in the sense of ${}^1\| \cdot \|_w$ for any $w > 0$. In fact, for any function g, be it $(F - f)$ itself or one of its derivatives,

$$\sup_{x_n \in \mathcal{D}_n} |g(x_n)| = \max_{\gamma \subset \mathcal{D}_n} \max_{x_n \in \gamma} |g(x_n)| \qquad [5.50]$$

and:

$$\int_{\mathcal{D}_n} |g(x_n)|^w dx_n \leq \mu(\mathcal{D}_n) \max_{x_n \in \mathcal{D}_n} |g(x_n)|^w \qquad [5.51]$$

where the $\mu(\cdot)$ is a measure defined over $\mathcal{D}_n$.

Thus, as long as the target function has a bounded Hessian, a piecewise affine model is sufficient to approximate its values and its derivatives, providing that $\hat{d}$ is small enough.

In principle, this property can also be reversed. In fact, if a bound on the Hessian of the target function were known, then an upper bound on the diameter of the affine regions (and thus a hint on their number) would be available, depending on the desired accuracy.

Regrettably, in practical cases this rarely happens and we are only guaranteed that decreasing the coarseness of the partition will eventually improve the identification.

As a final remark, note that even if the derivatives are discontinuous at region boundaries, an easy implication of the previous results ensures that as regions surrounding a given point shrink, then in that point the derivative of the approximation in any direction tends to the derivative of the target.

The consistency of the piecewise-polynomial approach to derivative approximation is therefore guaranteed, even along region boundaries.

5.4.4. *Model continuity and domain partitioning*

Since the model given by equation [5.41] is to be assumed continuous, $f(\cdot)$ is also continuous over the whole $\mathcal{D}_n$ (Fantuzzi *et al.* 2002).

In such a case, the parameter vectors are constrained to satisfy the following relation:

$$\lim_{\substack{x_n(t) \to \bar{x}_n \\ x_n(t) \in R_n^{(i')}}} f(x_n(t)) = \lim_{\substack{x_n(t) \to \bar{x}_n \\ x_n(t) \in R_n^{(i'')}}} f(x_n(t)) \qquad [5.52]$$

with $\bar{x}_n$ being an accumulation point for both $R_n^{(i')}$ and $R_n^{(i'')}$, that is, if:

$$[\bar{x}_n(t), 1]^T \, a^{(i')} = [\bar{x}_n(t), 1]^T \, a^{(i'')} \qquad [5.53]$$

The straightforward application of equation [5.53] to all of the accumulation points common to neighboring regions leads to an infinite number of constraints.

Moreover, the following result shows that the adoption of regions with straight borders guarantees that only a finite number of them are linearly independent.

Let $R_n^{(i',i'')}$ be the set of all of the accumulation points of two neighboring regions $R_n^{(i')}$ and $R_n^{(i'')}$. If $R_n^{(i',i'')}$ is convex, and p points $x_n^1(t), \ldots, x_n^p(t) \in R_n^{(i',i'')}$ exist for which equation [5.53] is satisfied, then equation [5.53] is also satisfied by any point $\bar{x}_n(t)$ of their convex hull.

This result can be verified as follows. In fact, if $\bar{x}_n(t)$ belongs to the convex hull of $x_n^1(t), \ldots, x_n^p(t)$ then p non-negative scalars $\delta_1, \ldots, \delta_p$ exist such that:

$$\sum_{k=1}^{p} \delta_k = 1 \qquad [5.54]$$

and:

$$\bar{x}_n(t) = \sum_{k=1}^{p} \delta_k x_n^k(t). \qquad [5.55]$$

Then the continuity constraints $[x_n^k, 1]^T(t)\ a^{(i')} - [x_n^k, 1]^T(t) a^{(i'')} = 0$ for $k = 1, 2, \ldots, p$, can be combined by means of equations [5.55] and [5.54] to obtain the result.

These previous results suggest that regions whose boundaries are convex polyhedra should be considered. In this case, in fact, continuity can be ensured by simply setting the value of the local models on the vertices of the boundaries only.

In this case, the previous result guarantees that the continuity constraints (one for each polyhedral vertex) can be collected in a finite matrix C_n such that:

$$C_n A_n = 0 \qquad [5.56]$$

being:

$$A_n = \begin{bmatrix} a^{(1)} \\ \vdots \\ a^{(M)} \end{bmatrix} \qquad [5.57]$$

In particular, it is undoubtedly convenient to 'triangulate' the domain $\mathcal{D}_n$, that is, to partition it into $2n$-dimensional simplexes.

Moreover, it is assumed that the triangulation is such that two simplexes are either disjointed, or have a whole k-dimensional boundary in common, with $k = 0, 1, \ldots, 2n - 1$.

In this way, the local affine model of equation [5.44] can be forced to assume given values at most in $2n + 1$ vertices of each simplex, which are affinely independent points.

If this point of view is adopted, the proposed approach depends on the availability of a systematic procedure to triangulate $2n$-dimensional domains. Moreover, if it is assumed that both inputs and outputs are confined within certain intervals, such a domain is actually a $2n$-dimensional hyper-rectangle, which can be triangulated by means of a *recursive* procedure.

Hence, given the triangulation of a $2(n - 1)$-dimensional hyper–rectangular domain, the procedure gives us the triangulation of a $2n$-dimensional hyper–rectangular domain by applying two conceptually identical steps.

This approach yields the same final triangulation as the non-recursive approach proposed in Rovatti *et al.* (1998) and thus, even if it is only asymptotically efficient in terms of the number of generated simplexes (Mara 1976; Cottle 1982; Sallee 1982, 1984; Haiman 1991), it benefits from the fast simplex searching feature that is highlighted in Rovatti *et al.* (1998).

Moreover, its statement, in recursive terms, helps to cope with the fact that at the beginning, the final model order n is not known. Instead, the order n is increased until satisfactory results are obtained. In this case, recursive triangulation allow the reuse of some of the work done for lower dimensional domains.

To begin with, it is assumed that $\mathcal{U} = \bigcup_i \mathcal{I}_i^{\mathcal{U}}$ and $\mathcal{Y} = \bigcup_i \mathcal{I}_i^{\mathcal{Y}}$, where the $\mathcal{I}_i^{\mathcal{U}}$ and $\mathcal{I}_i^{\mathcal{Y}}$ are intervals.

When $n = 1$, the natural partition of each rectangle is exploited, that is the Cartesian product $\mathcal{I}_{i'}^{\mathcal{U}} \times \mathcal{I}_{i''}^{\mathcal{Y}}$ in two triangles separated by one of the two diagonals.

Figure 5.4 shows an example of this partition when only one interval is present in both $\mathcal{Y}$ and $\mathcal{U}$ (Fantuzzi *et al.* 2002). Note that the regions $R_1^{(1)}$ and $R_1^{(2)}$ in Figure 5.4 have a triangular shape.

Then, it is assumed to have a triangulation of the $2n$-dimensional domain $\mathcal{U}^n \times \mathcal{Y}^n$ into simplexes. For any given simplex s of that triangulation and any given interval $\mathcal{I}_i^{\mathcal{Y}}$, the Cartesian product $p = s \times \mathcal{I}_i^{\mathcal{Y}}$ is considered.

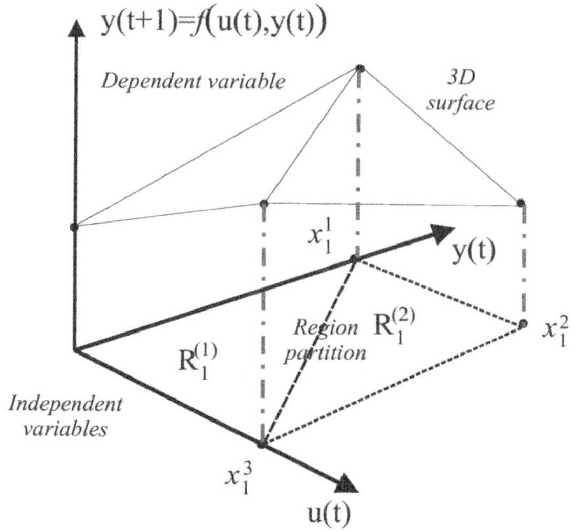

Figure 5.4. *Two–dimensional space partition example* $\mathcal{D}_1 = \mathcal{U} \times \mathcal{Y}$

Thus, a $(2n + 1)$-dimensional rectangular prism is obtained, whose faces are the two $2n$-dimensional simplexes s and s', the latter being the translation of s in the direction of the last axis, and for a length equal to the length of $\mathcal{I}_i^{\mathcal{Y}}$.

Now let $\xi_0, \ldots, \xi_{2n}$ be the vertices of s and $\xi'_0, \ldots, \xi'_{2n}$ be the vertices of s'. We triangulate the prism p by considering all of the $(2n + 1)$-dimensional simplexes that are the convex hulls of the $2n + 2$ vertices $\xi_0, \ldots, \xi_i, \xi'_i, \ldots, \xi'_{2n}$ for $i = 0, \ldots, 2n$. With this, the prism is triangulated into $2n + 1$ simplexes, all with the same volume.

If this is done for every simplex s of the triangulation of $\mathcal{U}^n \times \mathcal{Y}^n$ and for every interval $\mathcal{I}_i^{\mathcal{Y}}$, a triangulation of $\mathcal{U}^n \times \mathcal{Y}^{n+1}$ is obtained.

Most naturally, $(2n + 2)$-dimensional prisms are considered, which are the Cartesian products of a simplex of this last triangulation and an interval $\mathcal{I}_i^{\mathcal{U}}$, to obtain a triangulation of $\mathcal{U}^{n+1} \times \mathcal{Y}^{n+1}$.

Figure 5.5 shows how each of the steps for the triangulation of prisms works in the case of a three-dimensional domain.

5.4.5. *Local affine model estimation*

This section recalls the local prototype identification procedure for nonlinear system modeling proposed in Fantuzzi *et al.* (2002).

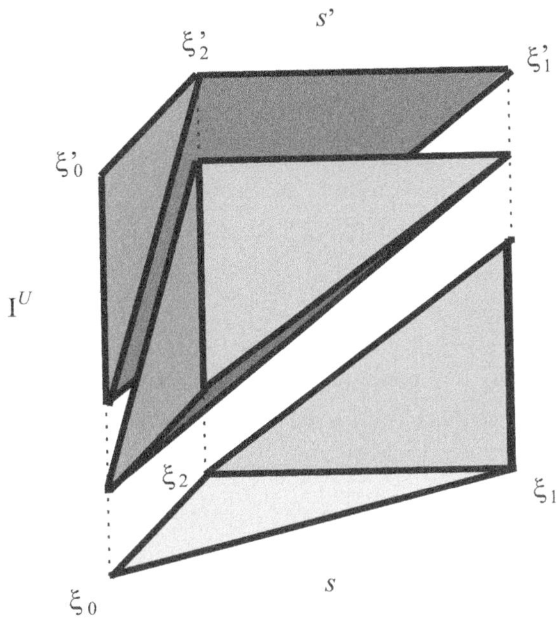

Figure 5.5. *Prism triangulation example*

The basic idea stems from the system identification with realistic measurements affected by errors (presented in section 5.3.2), slightly adapted to handle affine instead of homogeneous (linear) models.

It is assumed that the input-output measurements $u(t)$ and $y(t)$, $(t = 0, 1, \ldots, L_i)$ generated by a SISO system of the type from equation [5.42] are available.

In order to determine the order n and the parameters $a^{(i)}$ for the local model from equation [5.42] defined in region $R_n^{(i)}$, the following matrix is required:

$$X_k^{(i)} = \begin{bmatrix} y(k) & x_k^T(0) & 1 \\ y(k+1) & x_k^T(1) & 1 \\ \vdots & \vdots & \\ y(k+N_i-1) & x_k^T(N_i-1) & 1 \end{bmatrix}$$

$$\Sigma_k^{(i)} = \left(X_k^{(i)}\right)^T X_k^{(i)} \tag{5.58}$$

with $k + N_i - 1 \leq L_i$, and N_i is selected so that $k + N_i - 1$ is large enough to avoid unwanted linear dependence relationships due to limitations in the dimension of the vector spaces involved.

To determine the model order n in region $R_n^{(i)}$, it is possible to consider the sequence of increasing-dimension positive-definite or positive-semidefinite $\big((2k+2) \times (2k+2)\big)$ symmetric matrices:

$$\Sigma_2^{(i)}, \ \Sigma_3^{(i)}, \ \ldots \ \Sigma_k^{(i)}, \ \ldots \qquad [5.59]$$

testing their singularity as k increases. As soon as a singular matrix $\Sigma_k^{(i)}$ is found, then $n = k$, while the parameters $a^{(i)}$ describe the dependence relationship of the first vector of $\Sigma_n^{(i)}$ on the remaining ones as:

$$\Sigma_n^{(i)} \begin{bmatrix} -1 \\ a^{(i)} \end{bmatrix} = 0 \qquad [5.60]$$

It is worth observing that the vectors $x_n(0), x_n(1), \ldots, x_n(N_i - 1)$ in equation [5.58] must belong to the region $R_n^{(i)}$, according to the partition defined in the relation from equation [5.43]. Also note that in the presence of measurement errors, the procedure above for determining the order and model parameters would obviously be useless, since the matrices Σ_k would always be non-singular (positive-definite).

In order to solve the problem in a mathematical framework, it is necessary to characterize the measurement errors affecting the input-output data by suitable noise process models.

According to Frisch (1934); Kalman (1982b) and Beghelli *et al.* (1990), the following assumptions are made. The noise processes $\tilde{u}(t)$ and $\tilde{y}(t)$ are assumed to be additive on the input-output data $u^*(t)$ and $y^*(t)$ and region independent, so that:

$$u(t) = u^*(t) + \tilde{u}(t) \qquad [5.61]$$

$$y(t) = y^*(t) + \tilde{y}(t) \qquad [5.62]$$

Clearly, only $u(t)$ and $y(t)$ are available for the identification procedure, and moreover, every noise term $\tilde{u}(t)$ and $\tilde{y}(t)$ is modeled with a zero-mean white process and is supposed to be independent of every other term. These structures are the well-known Errors-In-Variables (EIV) descriptions.

Under these assumptions, $\bar{\tilde{\sigma}}_u$ and $\bar{\tilde{\sigma}}_y$ are the input and output noise variances respectively. The generic positive-definite matrix $\Sigma_k^{(i)}$ associated with the input-output noise-corrupted sequences can always be expressed as the sum of two terms $\Sigma_k^{(i)} = \Sigma_k^{*(i)} + \bar{\tilde{\Sigma}}_k$, where:

$$\bar{\tilde{\Sigma}}_k = \text{diag}[\bar{\tilde{\sigma}}_y I_{k+1}, \bar{\tilde{\sigma}}_u I_k, 0] \geq 0 \qquad [5.63]$$

Thus, also in this case, it is possible to determine the order and parameters of the model in region $R_n^{(i)}$ from the analysis of the sequence of increasing-dimension $\big((2k+2) \times (2k+2)\big)$ symmetric positive-definite matrices:

$$\Sigma_2^{(i)}, \ \Sigma_3^{(i)}, \ \ldots \ \Sigma_k^{(i)}, \ \ldots \qquad [5.64]$$

The solution to the above identification problem requires the computation of the unknown noise covariance values $\bar{\tilde{\sigma}}_u$ and $\bar{\tilde{\sigma}}_y$, which can be achieved by solving the following relation:

$$\Sigma_k^{*(i)} = \Sigma_k^{(i)} - \tilde{\Sigma}_k \geq 0. \qquad [5.65]$$

with respect to the variables $\tilde{\sigma}_u$, $\tilde{\sigma}_y$, where $\tilde{\Sigma}_k = \text{diag}[\tilde{\sigma}_y I_{k+1}, \tilde{\sigma}_u I_k, 0]$.

It is worth noting that the set of values of variables $\tilde{\sigma}_u, \tilde{\sigma}_y$ which make matrix $\Sigma_k^{*(i)}$ positive-semidefinite form a curve, as depicted in Figure 5.6.

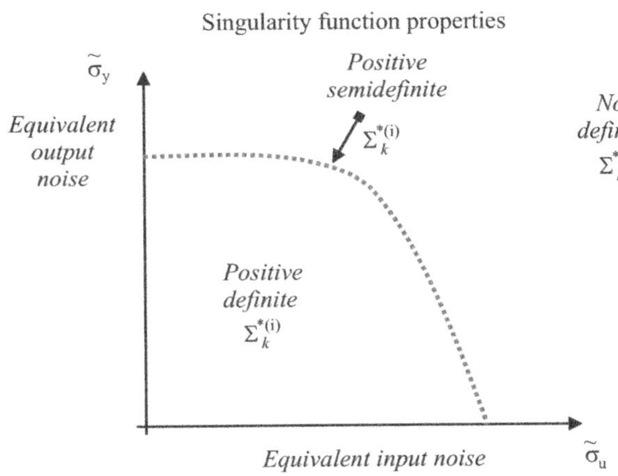

Figure 5.6. *Singularity curve example*

Unfortunately, the relation from equation [5.65] admits for any k, an infinite solution set describing a curve $\Gamma_k^{(i)}(\tilde{\sigma}_y, \tilde{\sigma}_u) = 0$ in the first orthant of the noise plane, whose concavity faces the origin. As an example, Beghelli *et al.* (1990) provide a constructive methodology to numerically compute this curve.

Since the determination of the system order requires the increasing values of k to be tested, it is relevant to analyze the behavior of the associated curves when k varies. As proved in Beghelli *et al.* (1990), the solution sets of condition [5.65] for different values of k are non-crossing curves in the noise plane $(\tilde{\sigma}_y,\, \tilde{\sigma}_u)$.

It is also important to observe that, since it is assumed that a system of the type given in equation [5.42] has generated the noiseless data, for $k \geq n$, all of the curves of the type given in equation [5.65] have at least one common point, that is, point $(\bar{\tilde{\sigma}}_u, \bar{\tilde{\sigma}}_y)$ corresponding to the true variances of the noise affecting the input and the output data.

The determination of a solution for the identification problem can thus start from the computation of this point in the noise space. This task can be achieved on the basis of the following properties.

With reference to the diagonal non-negative definite matrices $\bar{\tilde{\Sigma}}_k$, the following properties hold:

1) if $k < n$, the matrices $\Sigma_k^{*(i)}$ are positive-definite;

2) if $k > n$, the dimension of the null space of $\Sigma_k^{*(i)}$ and consequently, the number of eigenvalues equal to zero is $(k - n + 1)$;

3) for $k = n$, the matrix $\Sigma_k^{*(i)}$ is characterized by a linear dependence relation among its $2k + 2$ vectors, and the coefficients which link the first vector of $\Sigma_k^{*(i)}$ to the remaining ones are the parameters $a^{(i)}$, of the system [5.42], which has generated the noiseless sequences;

4) for $k \geq (n + 1)$, all of the $k - n + 1$ linear dependence relations among the vectors of the matrix Σ_k^* are characterized by the same $2n + 2$ coefficients $a^{(i)}$.

As an example, Figure 5.3 shows the above features for a system in the form from equation [5.42] and $n = 3$. The point marked by a circle corresponds to the input-output noise variance values $\bar{\tilde{\sigma}}_y$ and $\bar{\tilde{\sigma}}_u$ affecting the measurements. Figure 5.3 highlights that for $k = 3$ and $k = 4$ the curves share the common point $\bar{\tilde{\sigma}}_u, \bar{\tilde{\sigma}}_y$ (marked by a circle).

5.4.6. *Multiple-model estimation*

Section 5.4.5 discussed a procedure for the estimation of the noise variance values $\bar{\tilde{\sigma}}_u$ and $\bar{\tilde{\sigma}}_y$ and of the model order n, with respect to a particular region $\mathcal{A}_n^{(i)}$, as proposed in Fantuzzi *et al.* (2002).

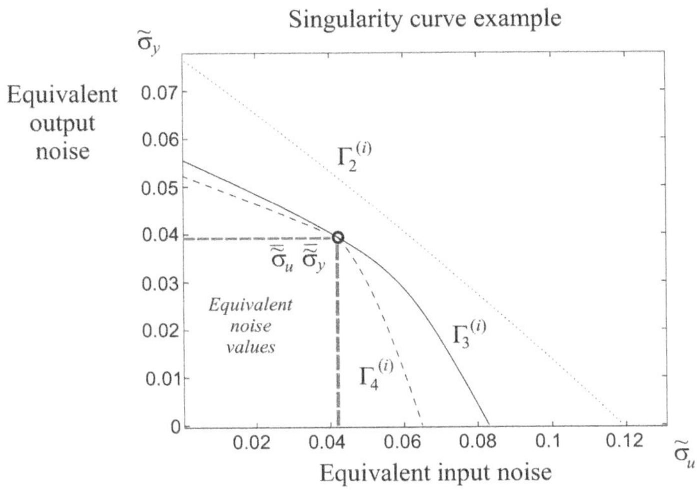

Figure 5.7. *Singularity curve example*

If the noise characteristics are common to all of the regions $\mathcal{A}_n^{(i)}$, since the physical nature of the process generating the noise is independent of the model structure and of the partition of $\mathcal{D}_n$, and all of the assumptions regarding the Frisch scheme are fulfilled, a common point $(\bar{\bar{\sigma}}_y, \bar{\bar{\sigma}}_u)$ in the noise plane exists for the singularity surfaces.

For a SISO model, in an two-dimensional space having $\bar{\bar{\sigma}}_u$ and $\bar{\bar{\sigma}}_y$ as coordinates, according to the previous sections, $\Gamma_k^{(i)}$ is a convex hypersurface belonging to the first orthant.

If the noise characteristics are common to all of the regions $\mathcal{A}_n^{(i)}$, as the physical nature of the process generating the noise is independent of the model structure and of the operating point conditions (i.e., the partition of $\mathcal{D}_n$ into regions) and all assumptions regarding the Frisch Scheme are fulfilled, for all the regions a common noise matrix $\bar{\bar{\Sigma}}_n$ exists that must satisfy:

$$\bar{\bar{\Sigma}}_n \in \bigcap_{i=1}^{M} \Gamma_k^{(i)} \qquad [5.66]$$

In this case, the right-hand side of [5.66] degenerates to a single point in the noise space. Therefore, the unique estimation of the noise process and its complete rejection is allowed.

The existence conditions of a unique common noise matrix from equation [5.66] for all of the regions can be found by considering the following results.

Let $\Sigma_j^{(i)}$ be obtained from $\Sigma^{(i)}$, deleting its jth row and column. If for any j', j'' there are two indices i', i'' such that:

$$\left(\det(\Sigma_{j'}^{(i')}) - \det(\Sigma_{j'}^{(i'')})\right)\left(\det(\Sigma_{j''}^{(i')}) - \det(\Sigma_{j''}^{(i'')})\right) \leq 0 \qquad [5.67]$$

and $M > m$, then the right-hand side of equation [5.66] degenerates into a single point.

Under these conditions, as an example, the singularity surfaces regarding two regions $R_n^{(i)}$ and $R_n^{(j)}$ for a model with order $n = 3$ are depicted in Figure 5.8. The curves share the common point $(\bar{\tilde{\sigma}}_u, \bar{\tilde{\sigma}}_y)$ representing the variances of the true noises which affect the data.

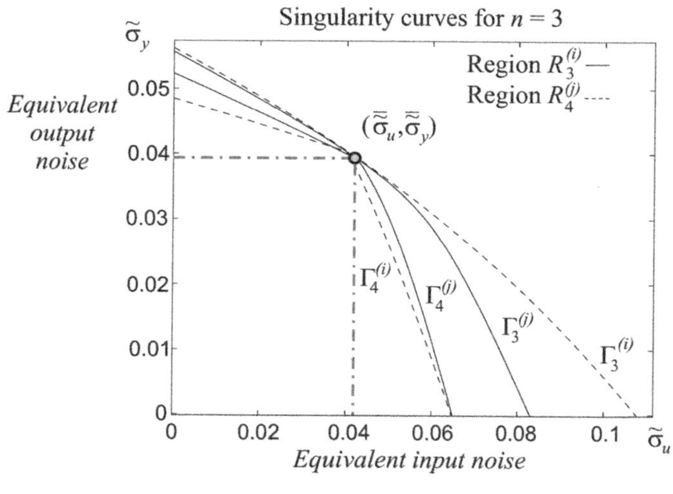

Figure 5.8. *Singularity curves example*

Once the order n has been determined, the parameters $a^{(i)}, i = 1, \ldots, M$ can be identified by solving the following relation:

$$(\Sigma_n^{(i)} - \bar{\tilde{\Sigma}}_n)\, a^{(i)} = 0 \quad \text{for } i = 1, \ldots, M \qquad [5.68]$$

The previous result can be fully applied when the assumptions behind the Frisch scheme are satisfied (independence between input-output sequences, additive noise, noise whiteness).

In real applications, these assumptions have to be relaxed, thus no common point can be determined among the surfaces $\Gamma_n^{(i)} = 0$ in the noise plane, and a unique solution to the identification problem can only be obtained by introducing a criterion to select a different noisy point for each region as a best approximation of the ideal case.

With reference to the identification of the system order n in the ith region $\mathcal{A}_n^{(i)}$, it must be noted that the $\Gamma_{n+1}^{(i)} = 0$ curve has a single point in common with the $\Gamma_n^{(i)} = 0$ curve in ideal conditions, which corresponds to a double singularity of the matrix $\Sigma_{n+1}^{*(i)}$.

In real cases, the order n can be computed, finding the point $(\tilde{\sigma}_u, \tilde{\sigma}_y) \in \Gamma_{n+1}^{(i)} = 0$ that makes $\Sigma_{n+1}^{*(i)}$ closer to the double singular condition (i.e., minimal eigenvalue equal to zero and the second minimum eigenvalue near to zero).

As n is unknown, increasing the system orders k must be tested, and the value of k associated with the minimum of the second eigenvalue of the matrix $\Sigma_{k+1}^{*(i)}$ corresponds to the order n. This criterion is consistent as it leads to the common point of the surfaces when the assumptions of the Frisch scheme are not violated.

Note that since the order n of the piecewise model [5.44] is region independent, it can be determined by choosing a k that fulfils the following inequality:

$$\max_{i=1,\ldots,M_k} \lambda_k^{(i)} < \epsilon \qquad [5.69]$$

when ϵ is an arbitrary positive constant and $\lambda_k^{(i)}$ is the minimal eigenvalue different from the zero of matrix $\Sigma_{k+1}^{*(i)}$.

This result leads to the derivation of the following algorithm for the selection of the model order, as addressed in Fantuzzi *et al.* (2002):

1) fix ϵ, k and M_k (k is the initial hypothesis on model order);

2) construct partitions $\{\mathcal{A}_k^{(1)}, \ldots, \mathcal{A}_k^{(M_k)}\}$;

3) cluster data into partitions;

4) compute matrices $\Sigma_{k+1}^{*(i)}$ from data clustered in region $\mathcal{A}_k^{(i)}$;

5) compute the test from equation [5.69]:

- *If success*: $n = k$, *exit*,

- *else* $k = k + 1$, *goto* 2.

Once the model order n is selected, the parameters $a^{(i)}, i = 1, \ldots, M$ cannot be computed from equation [5.68], because the surfaces $\Gamma_n^{(i)} = 0$ do not share the common point $(\bar{\tilde{\sigma}}_u, \bar{\tilde{\sigma}}_y)$.

In this case, for each region different input-output noise variances $(\bar{\tilde{\sigma}}_u^{(i)}, \bar{\tilde{\sigma}}_y^{(i)})$ must be considered and the relation [5.65] should be rewritten as:

$$\Sigma_n^{*(i)} = \Sigma_n^{(i)} - \tilde{\Sigma}_n^{(i)} \geq 0 \qquad [5.70]$$

where $\tilde{\Sigma}_n^{(i)} = \text{diag}[\bar{\tilde{\sigma}}_u^{(i)} I_{n+1}, \bar{\tilde{\sigma}}_y^{(i)} I_n, 0]$.

And $(\bar{\tilde{\sigma}}_u^{(i)}, \bar{\tilde{\sigma}}_y^{(i)})$ represent the variance values of the equivalent input and output additive noise in region $R_n^{(i)}$.

The values $(\bar{\tilde{\sigma}}_u^{(i)}, \bar{\tilde{\sigma}}_y^{(i)})$ can be computed by solving an optimization problem that minimizes both of the distances between $(\bar{\tilde{\sigma}}_u^{(i)}, \bar{\tilde{\sigma}}_y^{(i)})$ and $(\bar{\tilde{\sigma}}_u^{(j)}, \bar{\tilde{\sigma}}_y^{(j)})$, with $i \neq j$ and the continuity constraints proposed in equation [5.56]:

$$\begin{aligned} J\left((\bar{\tilde{\sigma}}_u^{(1)}, \bar{\tilde{\sigma}}_y^{(1)}), \ldots, (\bar{\tilde{\sigma}}_u^{(M)}, \bar{\tilde{\sigma}}_y^{(M)})\right) = d\left((\bar{\tilde{\sigma}}_u^{(1)}, \bar{\tilde{\sigma}}_y^{(1)}), \ldots, (\bar{\tilde{\sigma}}_u^{(M)}, \bar{\tilde{\sigma}}_y^{(M)})\right) + \\ + (C_n A_n)^T H C_n A_n \end{aligned} \qquad [5.71]$$

with H being a positive-definite weighting matrix and d a distance defined as:

$$\begin{aligned} & d\left((\bar{\tilde{\sigma}}_u^{(1)}, \bar{\tilde{\sigma}}_y^{(1)}), \ldots, (\bar{\tilde{\sigma}}_u^{(M)}, \bar{\tilde{\sigma}}_y^{(M)})\right) = \\ & = \sum_{i=1}^{M} \sum_{j=i+1}^{M} \sqrt{(\bar{\tilde{\sigma}}_u^{(i)} - \bar{\tilde{\sigma}}_u^{(j)})^2 + (\bar{\tilde{\sigma}}_y^{(i)} - \bar{\tilde{\sigma}}_y^{(j)})^2}. \end{aligned} \qquad [5.72]$$

and $s^{(i)} : q_i \in [0, 1] \mapsto (\tilde{\sigma}_u^{(i)}, \tilde{\sigma}_y^{(i)}) \in [0, \tilde{\sigma}_{u,Max}^{(i)}] \times [0, \tilde{\sigma}_{y,Max}^{(i)}]$ being a parameterization of the curve $\Gamma_n^{(i)}(\tilde{\sigma}_u^{(i)}, \tilde{\sigma}_y^{(i)}) = 0$.

The cost function can be interpreted as a sum of a 'measure' of the distance among points $(\bar{\tilde{\sigma}}_u^{(i)}, \bar{\tilde{\sigma}}_y^{(i)})$ in the noise plane and a positive function representing the degree of fulfillment of continuity constraints.

It is worth observing that the matrix A_n collects the parameters $a^{(i)}, i = 1, \ldots, M$ which depend on $(\bar{\tilde{\sigma}}_u^{(i)}, \bar{\tilde{\sigma}}_y^{(i)})$.

Now, it is considered the problem of determining the model order n. In the real case, the issue of item two is only approximately fulfilled, that is, for $k > n$, the null eigenvalue has algebraic multiplicity one, whereas the *second* minimum eigenvalue is very close to zero.

The minimization of the cost function from equation [5.71] can be computationally difficult, as it depends on $2M$ independent variables.

Therefore, in order to decrease the complexity of the problem, a common parameterization can be defined for all of the surfaces $\Gamma_n^{(i)}(\bar{\bar{\sigma}}_u^{(i)}, \bar{\bar{\sigma}}_y^{(i)}) = 0$ by introducing the polar coordinates:

$$\begin{cases} \bar{\bar{\sigma}}_u^{(i)} = \rho^{(i)} \cos \frac{\pi}{2} q \\ \bar{\bar{\sigma}}_y^{(i)} = \rho^{(i)} \sin \frac{\pi}{2} q \end{cases} \qquad [5.73]$$

where $\rho^{(i)}$ is determined so that $\Gamma_n^{(i)}(\rho^{(i)} \cos \frac{\pi}{2} q, \rho^{(i)} \sin \frac{\pi}{2} q) = 0$ and $q \in [0, 1]$.

The cost function has the form:

$$\begin{aligned} J(q) = d &\left((\bar{\bar{\sigma}}_u^{(1)}(q), \bar{\bar{\sigma}}_y^{(1)}(q)), \ldots, (\bar{\bar{\sigma}}_u^{(M)}(q), \bar{\bar{\sigma}}_y^{(M)}(q)) \right) + \\ &+ (C_n A_n)^T H C_n A_n. \end{aligned} \qquad [5.74]$$

On the basis of the previous remark, the following algorithm to solve the problem is proposed.

1) set constants $\epsilon > 0$, $h > 0$, $k = 0$ and H;

2) set $k = k + 1$;

3) compute $(\bar{\bar{\sigma}}_u^{(i)}, \bar{\bar{\sigma}}_y^{(i)})$ and A_k to solve the optimization problem from equation [5.71];

4) compute $\lambda_{min}^{(i)}$ as the *second* minimum eigenvalue of matrices $\Sigma_k^{*(i)}$ for all $i = 1, \ldots, M$;

5) *if* $\max_{i=1,\ldots,M} \lambda_k^{(i)} < \epsilon$ *then* set $n = k - 1$ and $A_n = A_{k-1}$ *else* go to step 2.

The parameterization chosen to simplify the minimization problem leads to consistent results. In fact, when the data are generated by a continuous piecewise affine dynamic system, with all assumptions regarding the Frisch scheme being fulfilled and noise being region-independent, the surfaces $\Gamma_n^{(i)} = 0$ share a common point in the noise plane. In these conditions, the cost function $J(q) = 0$ and the variances $(\bar{\bar{\sigma}}_u, \bar{\bar{\sigma}}_y)$ are identified exactly.

Finally, it is worth observing how once the parameter q minimizing the cost function from equation [5.74] is computed, the matrices $\tilde{\Sigma}_n^{(i)}$ can be constructed and the model parameter $a^{(i)}, i = 1, \ldots, M$ determined by means of relation:

$$(\Sigma_n^{(i)} - \tilde{\Sigma}_n^{(i)})a^{(i)} = 0 \quad \text{for } i = 1, \ldots, M. \qquad [5.75]$$

This completes the multiple-model identification procedure. Also note that this choice of parameterization is compatible with the ideal case in which a common point among the curves $\Gamma_n^{(i)}$ exists in the noise plane.

These solutions can be exploited for the identification of models that describe the behavior of real nonlinear processes using piecewise affine prototypes, and are then exploited for the generation of residual diagnostic signals (Fantuzzi *et al.* 2002).

5.5. Fuzzy data-driven approach to fault diagnosis

Since its introduction in 1965, fuzzy set theory has found applications in a wide variety of disciplines. The modeling and control of dynamic systems belong to the fields in which fuzzy set techniques have received considerable attention, not only from the scientific community but also from the industry. Many systems are not amenable to conventional modeling approaches due to the lack of precise, formal knowledge about the system, due to strongly nonlinear behavior, the high degree of uncertainty, or the time varying characteristics (Babuška 1998).

Fuzzy modeling, along with other related techniques, such as neural networks, have been recognized as powerful tools which can facilitate the effective development of models. One of the reasons for this is the capability of fuzzy systems to integrate information from different sources, such as physical laws, empirical models, or measurements and heuristics.

Fuzzy models can be seen as logical models which use "IF-THEN" rules to establish qualitative relationships among the variables in the model. Fuzzy sets serve as a smooth interface between the qualitative variables involved in the rules and the numerical data at the inputs and outputs of the model. The rule-based nature of fuzzy models allows the use of information expressed in the form of natural language statements and consequently makes the models transparent to interpretation and analysis. At the computational level, fuzzy models can be regarded as flexible mathematical structures, similar to neural networks, that can approximate a large class of complex nonlinear systems to a desired degree of accuracy.

Recently, a great deal of research activity has focused on the development of methods to build or update fuzzy models from numerical data, exploited for fault diagnosis. Many approaches are also based on neuro-fuzzy systems, which exploit the functional similarity between fuzzy reasoning systems and neural networks. This "marriage" of fuzzy systems and neural networks enables a more effective use of optimization techniques for building fuzzy systems, especially with regard to their approximation accuracy. However, the aspects related to the transparency and interpretation tend to receive considerably less attention. Consequently, most neuro-fuzzy models can be regarded as 'gray-box' models, which provide little insight to help understand the underlying process.

The approach exploited in this section focuses on the identification of transparent rule-based fuzzy models, which can accurately predict the quantities of interest, and at the same time provide insight into the system that generated the data. Attention is paid to the selection of appropriate model structures in terms of the dynamic properties, as well as the internal structure of the fuzzy rules (in particular, Takagi–Sugeno type) (Takagi and Sugeno 1985). From the system identification point of view, a fuzzy model is regarded as a composition of local affine sub models. Fuzzy sets naturally provide smooth transitions between the submodels, and enable the integration of various types of knowledge within a common framework.

In order to automatically generate fuzzy models from measurements, a comprehensive methodology is developed. This employs fuzzy clustering techniques to partition the available data into subsets characterized by a linear behavior. The relationships between the presented identification method and linear regression are exploited, allowing for the combination of fuzzy logic techniques with system identification tools.

Using the concepts of model-based fault detection, the design of a residual generator based on a fuzzy model of a nonlinear dynamic process is addressed. The orientation of the section is toward methodologies that in the author's experience, proved to be practically useful. The presentation reflects theoretical and practical issues in a balanced way, aiming at readership from the academic world and the industry. Application examples are given in this book, where selected realistic applications are presented in detail.

5.5.1. *Fuzzy model identification*

The term "fuzzy identification" usually refers to techniques and algorithms for constructing fuzzy models from data. Two main approaches to the integration of knowledge and data in a fuzzy model can be distinguished (Babuška 1998):

1) The expert knowledge expressed in a verbal form is translated into a collection of **IF-THEN** rules. In this way, a certain model structure is created. Parameters in this structure (membership functions, weights of the rules, etc.) can be fine-tuned

using input-output data. The particular tuning algorithms exploit the fact that at the computational level, a fuzzy model can be seen as a layered structure (network), similar to artificial neural networks, to which standard learning algorithms can be applied. This approach is usually called neuro-fuzzy modeling.

2) No prior knowledge about the system under study is initially used to formulate the rules, and a fuzzy model is constructed using numerical data only. It is expected that the extracted rules and membership functions can provide an *a posteriori* interpretation of the system's behavior. An expert can confront this information with his own knowledge, modify the rules, or supply new ones, and design additional experiments in order to obtain more informative data.

These two techniques, of course, can be combined, depending on the particular application. This section mainly focuses on the presentation of methods and algorithms for the second approach, that is, for the automated acquisition of fuzzy models from data.

It is believed that this technique is more useful in practice, as it can obviate the process of knowledge acquisition, which is a well-known bottleneck for the practical applications of knowledge-based systems (McGraw and Harbisson-Briggs 1989). Instead, the expert is invited to assume a more active role in model analysis and validation, which may lead to revealing new pieces of information, and may result in a kind of emergent knowledge acquisition.

To date, relatively little attention has been devoted to the identification of transparent fuzzy models from data. Most of the techniques reported in literature aim to obtain numerical models that simply fit the data with the best possible accuracy, without paying attention to the interpretation of the results (Takagi and Sugeno 1985; Sugeno and Kang 1988; Johansen 1996; Wang 1995). Many other identification techniques can be used for completely "gray-box" modeling, such as standard nonlinear regression (Seber and Wild 1989), spline techniques (de Boor 1978; Brown and Harris 1994), or neural networks (Hunt *et al.* 1992b). In many cases, a natural requirement is that a model not only accurately predicts the system's outputs, but also provides some insight into the working of the system. Such a model can be used for the given situation, but can also be more easily adapted to changing design parameters and operating conditions.

In this section, fuzzy models are viewed as a class of local modeling approaches, which attempt to solve a complex modeling problem by decomposing it into a number of simpler subproblems. The theory of fuzzy sets offers an excellent tool for representing the uncertainty associated with the decomposition task, providing smooth transitions between the individual local sub models, and integrating various types of knowledge within one common framework.

In particular, fuzzy logic is exploited to define a Takagi–Sugeno (TS) fuzzy model (Takagi and Sugeno 1985). The TS fuzzy model for nonlinear dynamic systems is described by a collection of local linear or affine submodels, each one approximating the system behavior around a single working point. The scheduling of the submodels is achieved through a smooth function of the system state, the behavior of which is defined using fuzzy set theory (Klir and Yuan 1995).

Recalling some previous considerations, it can be recognized that such a structure fits the definition of the *multiple–model*, as stated by Leontaritis and Billings (1985a, 1985b). In fact, the basic approach to fuzzy modeling is similar to that presented in section 5.4.1, in which a number of local models are designed and the estimate of outputs is given by a smooth (fuzzy) fusion of local outputs.

A large part of fuzzy modeling and identification algorithms (see Babuška and Verbruggen 1995; Babuška *et al.* 1997; Babuška 1998 and references therein) share a common two-step procedure, in which at first, the operating regions are determined using heuristics or data clustering techniques. Then, in the second stage, the identification of the parameters of each submodel is achieved using a least-squares algorithm or Frisch scheme.

From this perspective, fuzzy identification can be regarded as a search for a decomposition of a nonlinear system, which gives a desired balance between the complexity and the accuracy of the model, effectively exploring the fact that the complexity of systems is usually not uniform. Since it cannot be expected that sufficient prior knowledge is available concerning this decomposition, methods for the automated generation of the decomposition, primarily from system data, are developed. A suitable class of fuzzy clustering algorithms is used for this purpose.

5.5.2. *Takagi–Sugeno prototypes*

A fuzzy rule–based model suitable for the approximation of a large class of nonlinear systems was introduced by Takagi and Sugeno (1985).

In the TS fuzzy model (see Figure 5.9), the rule consequents are crisp functions of the model inputs:

$$R_i : \textbf{ IF } x(t) \text{ is } A_i \textbf{THEN } y_i = f_i\big(x(t)\big), \; i = 1, 2, \cdots, K \qquad [5.76]$$

where $x(t) \in \Re^p$ is the input (antecedent) variable and $y_i \in \Re$ is the output (consequent) variable. R_i denotes the ith rule, and K is the number of rules in the rule base. A_i is the antecedent fuzzy set of the ith rule, defined by a (multivariate) membership function:

$$\mu_{A_i}(x) : \Re^p \to [0, 1] \, . \qquad [5.77]$$

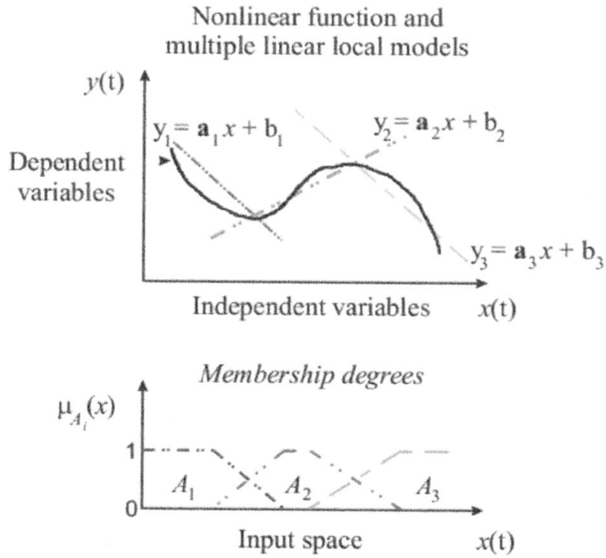

Figure 5.9. *Fuzzy model diagram*

As in the linguistic model, the antecedent proposition '$x(t)$ is A_i' is usually expressed as a logical combination of simple propositions with univariate fuzzy sets defined for the individual components of $x(t)$, often in the conjunctive form:

$$
\begin{aligned}
R_i: & \textbf{ IF } x_1 \text{ is } A_{il} \textbf{ and } x_2 \text{ is } A_{i2} \textbf{ and } \cdots \textbf{ and } x_p \text{ is } A_{ip} \\
& \textbf{THEN } y_i = f_i(x), i = 1, 2, \cdots, K.
\end{aligned}
\quad [5.78]
$$

The consequent functions f_i are typically chosen as instances of a suitable parameterized function, whose structure remains equal in all of the rules, and only the parameters vary. A simple and practically useful parameterization is the affine linear form:

$$y_i = a_i\, x + b_i, \quad [5.79]$$

where a_i is a parameter vector and b_i is a scalar offset. We refer to this model as an *affine TS model.* The consequents of the affine TS model are hyperplanes (p-dimensional linear subspaces) in $\Re^{p+1}$. The antecedent of each rule defines a (fuzzy) validity region for the corresponding affine consequent model. The global model is composed of a concatenation of the local models, and can be seen as a smoothed piecewise approximation of a nonlinear surface. Approximation properties of the affine TS model were investigated in Rovatti (1996).

A special case of the consequent function occurs when $b_i = 0, i = 1, \cdots, K$. Then the model is called a homogeneous TS model:

$$\textbf{IF } x \text{ is } A_i \textbf{ THEN } y_i = a_i x,\ i = 1, 2, \cdots, K. \quad [5.80]$$

This model has more limited approximation capabilities than the affine TS model (Fantuzzi and Rovatti 1996).

When $a_i = 0,\ i = 1, \cdots, K$, the consequents in model [5.79] are constant functions, and the *singleton model*, is obtained:

$$\textbf{IF } x \text{ is } A_i \textbf{ THEN } y_i = b_i,\ i = 1, 2, \cdots, K. \quad [5.81]$$

Before the output can be inferred, the degree of fulfillment of the antecedent denoted by $\beta_i(x)$ must be computed. For rules with multivariate antecedent fuzzy sets, given by equations [5.76] and [5.77], the degree of fulfillment is simply equal to the membership degree of the given input x, that is, $\beta_i = \mu_{A_i}(x)$. When logical connectives are used, the degree of fulfillment of the antecedent is computed as a combination of the membership degrees of the individual propositions using the fuzzy logic operators (Jager 1995; Babuška 1998).

In the Takagi–Sugeno model, the inference is reduced to a simple algebraic expression, similar to the fuzzy-mean defuzzification formula (Takagi and Sugeno 1985):

$$y = \frac{\sum_{i=1}^{K} \beta_i(x) y_i}{\sum_{i=1}^{K} \beta_i(x)} \quad [5.82]$$

By denoting the normalized degree of fulfillment:

$$\lambda_i(x) = \frac{\beta_i(x)}{\sum_{j=1}^{K} \beta_j(x)} \quad [5.83]$$

the affine TS model with a common consequent structure can be expressed as a pseudo-linear model with input-dependent parameters:

$$y = \Big(\sum_{i=1}^{K} \lambda_i(x) a_i^T \Big) x + \sum_{i=1}^{K} \lambda_i(x) b_i = a^T(x) x + b(x). \quad [5.84]$$

The parameters $a(x)$, $b(x)$ are convex linear combinations of the consequent parameters a_i and b_i, that is:

$$a(x) = \sum_{i=1}^{K} \lambda_i(x) a_i^T, \, b(x) = \sum_{i=1}^{K} \lambda_i(x) b_i. \quad [5.85]$$

Because of this property, a TS model can be regarded as a mapping from the antecedent (input) space to a convex region (polytope) in the space of the parameters of the quasi-linear system from equation [5.84].

Consider, for instance, a dynamic system described by the following TS rules:

$$\begin{aligned} R_i : \textbf{IF } & y(k) \text{ is } A_{il} \textbf{ and } y(k-1) \text{ is } A_{i2} \textbf{ and } \cdots y(k-n_y+1) \text{ is } A_{in_y} \\ & \textbf{and } u(k) \text{ is } B_{il} \textbf{ and } u(k-1) \text{ is } B_{i2} \textbf{ and } \cdots u(k-n_u+1) \text{ is } B_{in_u} \\ & \textbf{THEN } y(k+1) = \textstyle\sum_{j=1}^{n_y} y(k-j+1) + \sum_{j=1}^{n_u} u(k-j+1), \end{aligned} \quad [5.86]$$

where the consequents are linear ARX models (n_u and n_y are integers related to the order of the system).

5.5.3. *Data-driven Fuzzy modeling*

An effective approach to the identification of complex nonlinear systems is to partition the available data into subsets and approximate each subset by a simple model (Babuška and Verbruggen 1995). Fuzzy clustering can be used as a tool to obtain a partitioning of data where the transitions between the subsets are gradual rather than abrupt. This section gives an introduction to the basic concepts of fuzzy clustering (Babuška 1998).

The aim of this section is to explain clustering at a level necessary to understand the subsequent applications. For a more detailed treatment of the subject, the reader may refer to the classical monographs by Duda and Hart (1973), Bezdek (1981) and Jain and Dubes (1988). Another good overview can be found in Bezdek and Pal (1992) and the monograph by Backer (1995). The notation and terminology in this chapter follows Bezdek (1981).

5.5.4. *Clustering methods*

The aim of cluster analysis is the classification of objects according to similarities among them, and the organization of data into groups. Clustering techniques are among the unsupervised (learning) methods, since they do not use prior class identifiers. Most clustering algorithms do not rely on assumptions common to conventional statistical methods, such as the underlying statistical distribution of data, and therefore, they are useful in situations where little prior knowledge exists. The potential of clustering algorithms to reveal the underlying structures in data can be exploited, not only for classification and pattern recognition, but also for the reduction of complexity in modeling and optimization.

Clustering techniques can be applied to data which are typically observations of physical processes. Each observation consists of n measured variables, grouped into an n-dimensional column vector $z_k = [z_{lk}, \cdots, z_{nk}]^T$, $z_k \in \Re^n$. A set of N observations is denoted by $Z = \{z_k | k = 1, 2, \cdots, N\}$, and is represented as an $n \times N$ matrix:

$$Z = \begin{bmatrix} z_{ll} & z_{l2} & \cdots & z_{lN} \\ z_{2l} & z_{22} & \dots & z_{2N} \\ \vdots & \vdots & \vdots & \vdots \\ z_{nl} & z_{n2} & \cdots & z_{nN} \end{bmatrix} \qquad [5.87]$$

In the machine learning for pattern recognition terminology, the columns of this matrix are called patterns or objects, the rows are called the features or attributes, and Z is called the pattern or data matrix. The meaning of the columns and rows of Z depends on the context.

When clustering is applied to the modeling and identification of dynamic systems for fault diagnosis, the columns of Z contain samples of time signals, and the rows are, for instance, physical variables observed in the system under diagnosis. In order to represent the system dynamics, the past values of the variables are typically included in Z.

Generally, a cluster is a group of objects that are more similar to one another than to members of other clusters (Bezdek 1981; Jain and Dubes 1988). In metric spaces, similarity is often defined by means of a distance norm. Distance can be measured among the data vectors themselves, or as a distance from a data vector to some prototypical object of the cluster. The prototypes are usually not known beforehand, and are sought by the clustering algorithms, simultaneously with the partitioning of the data. The prototypes may be vectors of the same dimension as the data objects, but they can also be defined as geometrical objects, such as linear or nonlinear subspaces or functions. Data can reveal clusters of different geometrical shapes, sizes and densities.

Algorithms that can detect subspaces of the data space are of particular interest for identification and will be summarized in the following. Many clustering algorithms have been introduced in literature. Since clusters can formally be seen as subsets of the data set, one possible classification of clustering methods can be whether the subsets are fuzzy or crisp (hard). Hard clustering methods are based on classical set theory, and require that an object either belongs or does not belong to a cluster. *Fuzzy clustering* methods, however, allow the objects to belong to several clusters simultaneously, with different degrees of membership. In many situations, fuzzy clustering is more natural than hard clustering, as objects on the boundaries between several classes are not forced to fully belong to one of the classes, but are assigned membership degrees between O and 1, indicating their partial memberships.

Another classification of clustering techniques can be related to the algorithmic approach of the different techniques (Bezdek 1981). In particular, the class of clustering algorithms presented here exploits an objective function to measure the desirability of partitions. Nonlinear optimization algorithms are used to search for local extrema of the objective function.

Therefore, in the following, fuzzy clustering algorithms with objective functions will be presented. These methods lead to least-squares optimization, and hence, there are close relationships between clustering with fuzzy objective function and statistical regression and systems identification methods (Babuška 1998). In more detail, the clustering algorithm presented in this section is based on the optimization of the basic *c-means* objective function and it is known as a *fuzzy c-means clustering* algorithm (Dunn 1974).

5.5.5. *Fuzzy c-means clustering algorithms*

This fuzzy clustering algorithm is based on the minimization of the fuzzy functional formulated as (Bezdek 1981):

$$J(Z;U,V) = \sum_{i=1}^{c}\sum_{k=1}^{N}(\mu_{ik})^m \| z_k - v_i \|^2 \qquad [5.88]$$

where:

$$U = [\mu_{ik}] \in M_{f_c} \qquad [5.89]$$

is a fuzzy partition matrix of Z:

$$V = [v_1, v_2, \ldots, v_c], v_i \in \Re^n \qquad [5.90]$$

is a vector of cluster prototypes (centers), which have to be determined:

$$D^2_{ikA} = \| z_k - v_i \|_A = (z_k - v_i)^T A (z_k - v_i) \qquad [5.91]$$

is a squared distance norm, and:

$$m \in [1, \infty) \qquad [5.92]$$

is a weighting exponent which determines the fuzziness of the resulting clusters. The measure of dissimilarity in equation [5.88] is the squared distance between each data point z_k and the cluster prototype v_i. This distance is weighted by the power of the membership degree of that point $(\mu_{ik})^m$. The value of the cost function from equation [5.88] can be seen as a measure of the total variance of z_k from v_i.

The Fuzzy c–Means (FCM) algorithm consists of the minimization of the c–means functional from equation [5.88]. It represents a nonlinear optimization problem that can be solved by using a variety of available methods (Babu and Murty 1994; Bezdek *et al.* 1987; DeSarbo 1982).

The most popular method exploits the first-order conditions for the stationary points from equation [5.88]. They can be found by adjoining the constraint of the fuzzy partition (Babuška 1998):

$$\sum_{i=1}^{c} \mu_{ik} = 1 \, , 1 \leq k \leq N \qquad [5.93]$$

to J by means of Lagrange multipliers:

$$\bar{J}\,(Z; U, V, \lambda) = \sum_{i=1}^{c} \sum_{k=1}^{N} (\mu_{ik})^m D_{ikA}^2 + \sum_{k=1}^{N} \left[\sum_{i=1}^{c} \mu_{ik} - 1 \right], \qquad [5.94]$$

and by setting the gradients of $\bar{J}$ with respect to U, V and λ to zero. If $D_{ikA}^2 > 0$, $\forall i, k$ and $m > 1$, then $(U, V) \in M_{f_c} \times \Re^{n \times c}$ may only minimize function [5.88] if:

$$\mu_{ik} = \frac{1}{\sum_{j=1}^{c} \left(D_{ikA}^2 / D_{ikA}^2 \right)^{2/(m-1)}} \, , 1 \leq i \leq c \, , 1 \leq k \leq N \qquad [5.95]$$

and:

$$v_i = \frac{\sum_{k=1}^{N} (\mu_{ik})^m z_k}{\sum_{k=1}^{N} (\mu_{ik})^m} \qquad [5.96]$$

This solution also satisfies the fuzzy partition constraints defined by equations [5.97] and [5.98]:

$$\mu_{ik} \in [0,\ 1]\ , 1 \leq i \leq c\, , 1 \leq k \leq N \qquad [5.97]$$

and:

$$0 < \sum_{k=1}^{N} \mu_{ik} < N\, , 1 \leq i \leq c. \qquad [5.98]$$

It is worth noting that equation [5.96] gives v_i as the weighted mean of the data items that belong to a cluster, where the weights are the membership degrees. That is why the algorithm is called 'c-means'.

The alternating optimization scheme used by the FCM algorithm loops through the estimates of $U^{(l)}$ and $V^{(l)}$ at the step l. The termination error is usually given by choosing a value of ϵ so that $\max_{ik}(|\mu_{ik}^{(l)} - \mu_{ik}^{(l-1)}|) < \epsilon$. The convergence of the algorithm has been proven by Bezdek (1980).

The number of clusters c is the most important parameter, in the sense that the remaining parameters have secondary effects on U, compared with the effects of the number of clusters. The choice of the number of clusters is discussed in the following.

5.5.6. *Gustafson–Kessel clustering algorithm*

A family of algorithms can be derived from the basic FCM scheme and the most important will be presented in this section: the Gustafson–Kessel algorithm.

Gustafson and Kessel (1979) (GK) extended the standard fuzzy c-means algorithm by employing an adaptive distance norm, in order to detect clusters of different geometrical shapes in one data set.

Each cluster has its own norm-inducing matrix A_i, which yields the following norm:

$$D_{ikA_i}^2 = (z_k - v_i)^T A_i (z_k - v_i). \qquad [5.99]$$

The matrices A_i are used as optimization variables in the c-means functional, thus allowing each cluster to adapt the distance norm to the local topological structure of the data. Let A denote a c-tuple of matrices: $A = (A_l,\ A_2,\ \cdots,\ A_c)$. The objective functional of the GK algorithm is defined by:

$$J(Z;U,V,A) = \sum_{i=1}^{c}\sum_{k=1}^{N}(\mu_{ik})^m D_{ikA_i}^2 \qquad [5.100]$$

where $U \in M_{f_c}$, $V \in \Re^{n\times c}$ and $m > 1$. The solutions:

$$(U,V,A) = \text{arg min}_{M_{f_c}\times\Re^{n\times c}\times \text{PD}^n} J(Z;U,V,A), \qquad [5.101]$$

are stationary points of J, where PD^n denotes a space of $n \times n$ positive-definite matrices. For a fixed A, the conditions of equations [5.95] and [5.96] can be directly applied. However, the objective function from equation [5.100] cannot be directly minimized with respect to Ai, since it is linear in Ai. J can be made as small as desired by making Ai less positive-definite. To obtain a feasible solution, Ai must be constrained in some way. The usual way of accomplishing this is to constrain the determinant of Ai. Allowing the matrix Ai to vary, with its determinant fixed, corresponds to optimizing the cluster's shape, while its volume remains constant:

$$|A_i| = \rho_i\ ,\ \rho_i > 0\ ,\ \forall\, i \qquad [5.102]$$

Using the Lagrange multiplier method, the following expression for Ai is obtained:

$$A_i = [\rho_i \det(F_i)]^{1/n} F_i^{-1}, \qquad [5.103]$$

where F_i is the fuzzy covariance matrix of the ith cluster defined by:

$$F_i = \frac{\sum_{k=1}^{N}(\mu_{ik})^m (z_k - v_i)(z_k - v_i)}{\sum_{k=1}^{N}(\mu_{ik})^m} \qquad [5.104]$$

It is worth noting that without any prior knowledge, the cluster volumes ρ_i are simply fixed at 1 for each cluster. A drawback of the GK algorithm is that due to the constraint defined by equation [5.102], it can only find clusters of approximately equal volumes.

The eigenstructure of the cluster covariance matrix provides information about the shape and orientation of the cluster. The ratio of the lengths of the cluster's hyperellipsoid axes is given by the ratio of the square roots of the eigenvalues of F_i. The directions of the axes are given by the eigenvectors of F_i (Babuška 1998). The linear subspaces of the data space are represented by flat hyperellipsoids, that can be seen as hyperplanes. The eigenvector corresponding to the smallest eigenvalue determines the normal to the hyperplane, and can be used to compute optimal local linear models from the covariance matrix (Babuška 1998).

5.5.7. *Optimal number of clusters*

When clustering real data without any *a priori* information about the data structure, one usually has to make assumptions about the number of underlying subgroups (clusters) c in the data. The chosen clustering algorithm then searches for c clusters, regardless of whether they are really present in the data or not. Two main approaches to determining the appropriate number of clusters in data can be distinguished by (Babuška 1998):

1) Clustering data for different values of c, and using validity measures to assess the goodness of the obtained partitions. Different scalar validity measures have been proposed in literature.

2) Starting with a sufficiently large number of clusters, and successively reducing this number by merging clusters that are similar (compatible) with respect to some predefined criteria.

5.6. Fuzzy model identification

This section addresses the decomposition of a nonlinear identification problem into a set of locally linear models by means of product space fuzzy clustering. The identification procedure is first outlined and the structure selection and the choice of regressors in the modeling of dynamic systems are therefore discussed. Moreover, the principle of identification of nonlinear systems by product space clustering is of particular interest.

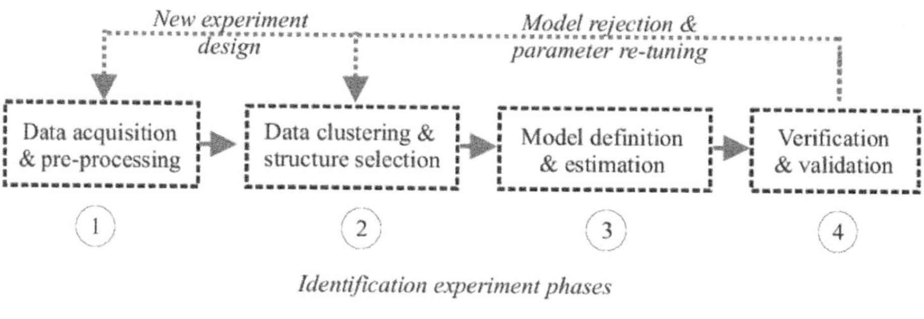

Figure 5.10. *Fuzzy clustering phases*

Figure 5.10 outlines the individual steps of the iterative identification procedure. As in a typical identification session, some of the steps may be repeated for the different choices of the various parameters. The different steps and the meaning of the blocks in Figure 5.10 are summarized below.

1) Data collection: this is an important initial step for any identification method, since it determines the information content of the identification data set. Although the choice of the excitation signal may be problem dependent, the input data should preferably excite the system in the entire range of the considered variables, both in amplitude and in frequency. Moreover, the choice of a suitable sampling period, the design of (anti–aliasing) filters, the duration of the experiments, etc., are other important issues of the experiment design (Norton 1986; Ljung 1999).

2) Structure selection: the purpose of this step is to determine the relevant input and output variables with respect to the aim of the modeling experiment. When identifying dynamic systems, the structure and the order of the model dynamics must be chosen. Structure selection allows us to translate the identification of a dynamic system into a regression problem that can be solved in a quasi-static manner. The structure can be selected in an automated way by comparing different candidate structures in terms of some performance measures. In most cases, a reasonable choice can be made by the user.

3) Data clustering: structure selection leads to a nonlinear static regression problem, which is then approximated by a collection of local linear submodels. The location and the parameters of the sub models are found by partitioning the available data into clusters. Each of the clusters defines a fuzzy region in which the system can be approximated locally by an affine submodel. Moreover, an appropriate number of clusters can be found. This stage can require several repetitions of step 3 for a different number of clusters.

4) TS model: fuzzy clustering divides the available data into groups, in which local affine relations exist between the inputs and the output. In order to obtain a model suitable for prediction, the TS fuzzy model parameters are identified from the available fuzzy partition matrix and from the cluster prototypes. Therefore, the membership functions and other parameters that constitute the fuzzy model are extracted in an automated way.

5) Validation: by means of validation, the final model is either accepted as appropriate for the given purpose, or it is rejected. In the latter case, some steps of the identification loop shown in Figure 5.10 may be repeated with a different setting, as this is usual in other approaches to linear and nonlinear system identification (Norton 1986; Ljung 1999). In addition to the usual numerical validation by means of simulation, the interpretation of fuzzy models plays an important role in the validation step. The coverage of the input space by the rules can be analyzed, and, for an incomplete rule base, additional rules can be provided based on prior knowledge, local linearization, or first-principle models.

In the fuzzy modeling and identification procedure, the problem of structure estimation can consists of several subproblems:

1) Input and output variable selection: although most identification methods assume that the input and output variables of the process are known (Ljung 1999; Norton 1986), in reality, especially for multivariable and closed-loop systems, it is often not clear which variables should be considered as the model inputs. The selection of the input and output variables is based on the aim of the modeling and identification experiment. This stage can also be partially automated. Several candidate models with different input variables can be compared in terms of some performance measures, and then the best one is selected (Guidorzi 1975, 1981).

2) System's order representation: a common approach is to transform the identification of a dynamic system into a static regression problem (Billings and Voon 1983a; Leontaritis and Billings 1985a, 1985b; Chen and Billings 1989; Sjöberg *et al.* 1995). The choice of this particular transformation is usually based on a combination of *a priori* knowledge with intuition, insight, and understanding of the process behavior. Physical modeling of the relationships and physical laws can guide the selection of the relevant variables, and of the model's order. This transformation can be regarded as a mapping from the domain of time signals into a space of variables that fully determine the state of the system. These variables are called the regressors. The system's behavior can be predicted by means of a static mapping from the space of regressors to the space of the model output (regressand). The choice of the regressors is a crucial step, as an inappropriate choice may lead to useless models.

3) Fuzzy model complexity: this choice is related to the number of rules in the model. In the fuzzy modeling literature, the term 'structure identification' is often applied to this step (Sugeno and Kang 1988). When fuzzy clustering is applied to identify fuzzy models from data, the number of clusters is the primary parameter that must be chosen. It should be noted that the complexity of the final model is also related to the particular type of model used. The ways in which these models approximate functions differ, which means that the model complexity needed to achieve a required level of accuracy may differ. In practice, there is usually a trade-off between the accuracy and complexity of the model.

5.6.1. *Nonlinear model identification*

Fuzzy systems are general function approximators and can therefore be applied to general nonlinear regression problems (Wang 1992; Kosko 1994; Zeng and Singh 1996). Nonlinear regression is the modeling of the static dependence of a response variable, called the regressand, $y \in Y \subset \Re$ on the regression vector $x = [x_l, \cdots, x_p]^T$ over some domain $X \subset \Re^p$.

The elements of the regression vector will be called the regressors and the domain X, the regressor space. The system that generated the data is presumed to be described by:

$$y \approx f(x) \qquad [5.105]$$

The deterministic function $f(\cdot)$ captures the dependence of y on x, and $\approx$ reflects the fact that y will not be an exact function of x.

As an example, the dynamics of the system under observation can be described by an EE model (Kalman 1982b, 1990):

$$y = f(x) + \varepsilon \qquad [5.106]$$

where ε can represent the error term.

The aim of regression is to use the data to construct a function $F(x)$ that can serve as a reasonable approximation of $f(x)$, not only for the given data, but over the entire domain X. The definition of 'reasonable approximation' depends on the purpose for which the model is constructed. If the aim of modeling is to obtain predictions of y, accuracy is the most relevant criterion. Lack of accuracy is usually defined by the integral error:

$$I = \int_X \| f(x) - F(x) \| \, \mathrm{d}x, \qquad [5.107]$$

over the entire domain X. In general, this error cannot be computed, since the value of f is only known at the available data points. Therefore, the average prediction error over the available data is often used:

$$J = \frac{1}{N} \sum_{i=1}^{N} \| f(x_i) - F(x_i) \| , \qquad [5.108]$$

where N denotes the number of data samples. The attainment of the minimum of I in equation [5.107] implies the best model possible with the selected structure. This is, however, not the case with the criterion J, which only guarantees that the model fits the available data with the least-squares error. Hence, a separate validation step is necessary, in order to assess the goodness of the model over the entire region of interest X.

Apart from accurate predictions, the goal may be to obtain a model that can be used to analyze and understand the properties of the real system that generated the data. A strong potential of fuzzy models is that they describe systems as a collection of simple local submodels that are expressed as rules. Rules can also be combined with analytical models, such as the local linear models in the TS structure described in section 5.5.2.

There are a number of possibilities for the choice of regressors in nonlinear black-box identification. Since extensive literature is available on this topic

(Leontaritis and Billings 1985a, 1985b; Chen and Billings 1989; Sjöberg *et al.* 1995), only a brief review is given here with respect to the use of the particular structures for clustering-based identification.

The NARX (Nonlinear AutoRegressive with eXogenous input) model is frequently used with many nonlinear identification methods, such as neural networks (Hunt *et al.* 1992a), radial basis functions (Chen *et al.* 1991), CMAC (Brown and Harris 1994), fuzzy (Lee *et al.* 1994) and neuro-fuzzy models. The NARX model establishes a relation between the past input-output data and the predicted output:

$$\hat{y}(k+1) = F\big(y(k), \cdots, y(k-n_y+1), u(k), \cdots, u(k-n_u+1)\big), \qquad [5.109]$$

where k denotes discrete time samples, n_u and n_y are integers related to the system's order, and F denotes a fuzzy model. In the NARX model, the regression vector is a collection of a finite number of past inputs and outputs, $x(k) = [y(k), \cdots, y(k-n_y+1), u(k), \cdots, u(k-n_u+1)]^T$. The regressand is the predicted output $\hat{y}(k+1)$. Hence, from a set of observed inputs and outputs of an unknown dynamic system, the function $F(\cdot)$ from equation [5.109] can be approximated by using static nonlinear regression.

The identification of the input-output model structure corresponds to the choice of the model type and of the related structural parameters n_y, n_u. *A priori* knowledge is typically used to make a first guess of the range of these parameters, and a structure is then sought within this range that minimizes a certain criterion. The most straightforward approach is to use the mean–square prediction error directly:

$$J = \frac{1}{N}\sum^{N} i = 1\big(y(i) - \hat{y}(i)\big)^2 \qquad [5.110]$$

evaluated on a different data set from that used to identify the system, in order to avoid fitting the noise.

Since linear identification techniques are much simpler and numerically more robust than nonlinear methods, it is usually worthwhile to start with a linear model to determine the structure. A variety of tools can be used, such as the singular value and the coherence tests (Verhaegen and Dewilde 1992) or the information-theoretic criteria (Akaike 1974; Rissanen 1978). The structure of the best linear model is then used as a starting point for nonlinear modeling. Also, cluster validity measures can guide the selection of the model's order and the number of clusters within the given structure.

It has been assumed that y is a scalar, that is, the system under study is a MISO system. With input-output models, MIMO systems can be represented in two ways:

the function F is a vector-valued function, or the MIMO system is decomposed into a set of coupled MISO systems. While the former approach is typically used with neural networks, in fuzzy modeling the decomposition approach is mostly adopted. The reason is that the it is more flexible if each output is associated with a different sort of nonlinearity. One output may contain a complex nonlinearity in a region, while another output may be linear in the same region. By decomposing the MIMO mapping into several MISO mappings, the number of membership functions and rules can be reduced.

5.6.2. *Product space clustering identification*

The principle of identification by product space clustering is to approximate a nonlinear regression problem, by decomposing it into several local linear subproblems. This approach has a number of advantages in comparison with global nonlinear models, such as neural networks. The model structure is easy to understand and interpret, both qualitatively and quantitatively. Various types of knowledge can be integrated into the model, including empirical knowledge, measured data and available mathematical models. In addition, the approach has computational advantages and lends itself to straightforward adaptive and learning algorithms (Murray-Smith and Johansen 1997).

Fuzzy clustering is applied in the product space of the regressors and the regressand: $X \times Y$. Let X denote the matrix in $\Re^{N \times p}$, having the regression vectors x_k^T in its rows, and let y denote the column vector in $\Re^N$, containing the regressands y_k:

$$X = \begin{bmatrix} x_1^T \\ \vdots \\ x_N^T \end{bmatrix}, y = \begin{bmatrix} y_1 \\ \vdots \\ y_N \end{bmatrix} \quad [5.111]$$

where N denotes the number of data samples and p is the dimension of the regression vector. For an input-output model of a dynamic system, the matrix X contains shifted versions of the input and output data.

The decomposition of a global nonlinear mapping into a set of locally linear models is based on a geometrical interpretation of the regression problem. The unknown nonlinear function $y = f(x)$ represents a nonlinear surface or hypersurface in the product space: $(X \times Y) \subset \Re^{p+1}$. This surface is called the regression surface.

The available data represent a sample from the regression surface. By clustering the data, local linear models can be found that approximate the regression surface in an optimal way. The set of data to be clustered, denoted Z, is constructed by concatenating the regressor data matrix X and the regressand vector y:

$$Z^T = [X, y] \qquad [5.112]$$

This data set is a subset of the Cartesian product space $X \times Y$ defined by the nonlinear functional relation from equation [5.105]:

$$Z \subset X \times y \text{ such that } y \approx f(X) \qquad [5.113]$$

The data set Z is partitioned into fuzzy subsets by applying fuzzy clustering algorithms capable of detecting linear substructures in data, according to section 5.5.3.

The membership of the data samples in the clusters is described by the fuzzy partition matrix. Each cluster is characterized by its center and covariance matrix, which represents the variance of the data in the cluster. The fuzzy clustering algorithm $\mathcal{C}$ can be regarded as a mapping $\mathcal{C} : (Z \times \mathbb{N}) \rightarrow (M_{f_c} \times \Re^{n \times c} \times \mathrm{PD}^n)$:

$$(U, V, F) = \mathcal{C}(Z, c; U^o, m, \epsilon), \qquad [5.114]$$

where c is the number of clusters, U^o is the initial partition matrix, while m and ϵ are the parameters of the clustering algorithm, according to section 5.5.3. The partition matrix U contains the membership degrees of the data points in the clusters with prototypes V.

It is worth noting that the cluster covariance matrix F_i contains information about the shape and orientation of the ith cluster. The jth eigenvalue and the jth unit eigenvector of F_i are denoted λ_{ij} and ϕ_{ij}, respectively (Babuška 1998).

The eigenvalues of the cluster covariance matrix F_i are arranged such that:

$$\lambda_{i1} \geq \lambda_{i2} \geq \cdots \geq \lambda_{in} \qquad [5.115]$$

and the eigenvectors are labeled accordingly. The eigenvectors ϕ_{i1} to $\phi_{i(n-1)}$ span the ith cluster linear subspace and the nth eigenvector ϕ_{in} is the normal to this linear subspace. Since λ_{in} is the smallest eigenvalue, ϕ_{in} is therefore called the smallest eigenvector.

When the intrinsic dimension of the data is p, λ_{in} is in orders of magnitude smaller than the remaining eigenvalues. This means that the n-dimensional data can be locally represented by a linear combination of $n - 1$ variables. Hence, the proportions between the eigenvalues can be used to check whether an appropriate structure has been chosen. If all of the eigenvalues are in the same order of magnitude, then no functional relationship between the regressors and the regressand has been detected (the chosen structure may not be rich enough). However, when several relatively small eigenvalues are found, the regression problem may be (locally) of a lower dimension than was assumed, or the data might not be rich enough (Babuška 1998).

5.6.3. *Fuzzy clustering model identification*

As shown in the previous sections, fuzzy clustering algorithms can be used to approximate a set of data by local linear models. Each of these models is represented by a fuzzy subset in the data set available for identification. In order to obtain a model useful for prediction, an additional step must be applied to generate a model independent of the identification data. Such a model can either be represented as a rule base or as a fuzzy relation (Babuška 1998). This section reviews algorithms for constructing fuzzy TS models from the fuzzy partitions obtained by product space clustering. In particular, the construction of Takagi–Sugeno models is addressed and methods for generating the antecedent membership functions and estimating the consequent parameters are presented below.

Each cluster obtained by product space clustering of the identification data set can be regarded as a local linear approximation of the regression hypersurface. The global model can be conveniently represented as a set of affine Takagi–Sugeno rules:

$$R_i : \textbf{IF } x \text{ is } A_i \textbf{ THEN } y_i = a_i^T x + b_i \, , \; i = 1, 2, \cdots, K. \qquad [5.116]$$

The antecedent fuzzy sets A_i can be computed analytically in the antecedent product space, or can be extracted from the fuzzy partition matrix. The *consequent parameters* a_i and b_i are estimated from the data using the methods presented in section 5.3 (i.e. Frisch scheme procedure), or they can be extracted from the eigenstructure of the cluster covariance matrices.

5.6.4. *Antecedent membership function estimation*

The antecedent membership functions can be obtained by projecting the fuzzy partition onto the antecedent variables, or by computing the membership degrees directly in the product space of the antecedent variables. These two methods are described in the following, but the second one will be exploited for the identification of the TS models by means of the procedure developed in Babuška (1998, 2000).

– Antecedent Membership Function Estimation by Projection: the principle of this method consists of projecting the multi-dimensional fuzzy sets defined pointwise in the rows of the partition matrix U onto the individual antecedent variables of the rules. These variables can be the original regression variables, in which case the projection is an orthogonal projection of the data. New, transformed antecedent variables can be obtained by means of eigenvector projection, using the p largest eigenvectors of the cluster covariance matrices. The eigenvector projection is useful for clusters which are opaque to the axis of the regression space, and cannot be represented by axis-orthogonal projection with a sufficient accuracy.

By projecting the ith row μ_i of the fuzzy partition matrix U onto the antecedent variable x_j, a point–wise definition of the fuzzy set A_{ij} is obtained. In order to obtain a prediction model, the antecedent membership functions must be expressed in a form that allows the computation of the membership degrees, and also for input data not contained in the data set Z. This is achieved by approximating the point-wise defined membership function by some suitable parametric function.

The piece-wise exponential membership functions proved to be suitable for the accurate representation of the actual cluster shape. This function is fitted to the envelope of the projected data by numerically optimizing its parameters. An advantage of this method over the *multi-dimensional membership functions*, summarized below, is that the projected membership functions can always be approximated, such that convex fuzzy sets are obtained. Moreover, asymmetric membership functions can be used to reflect the actual partition of the considered nonlinear regression problem.

– Multi-dimensional Antecedent Membership Functions: in this second method, the antecedent membership functions are represented analytically, by computing an inverse of the distance from the cluster prototype. The membership degree is computed directly for the entire input vector (without the decomposition). The antecedents of the TS rules are simple propositions with multi-dimensional fuzzy sets given by equation [5.116], and $\beta_i(x) = \mu_{A_i}(x)$.

Note that $F^x = [f_{ij}]$, $1 \leq i, j \leq p$ includes all but the last column of the cluster covariance matrix. The corresponding norm-inducing matrix is given by:

$$A_i^x = [\rho_i \det(F_i^x)]^{1/p} (F_i^x)^{-1}. \qquad [5.117]$$

$v_i^x = [v_{1i}, \cdots, v_{pi}]^T$ denote the projection of the cluster center onto X. The norm:

$$D_{A_i^x}(x, v_i^x) = (x - v_i^x)^T A_i^x (x - v_i^x) \qquad [5.118]$$

measures the distance of the antecedent vector x from the projection of the cluster centre v_i^x. To transform the distance into the membership degree, two methods can be applied, which are described below.

One possibility is to use the formula employed by the clustering algorithms:

$$\beta_i(x) = \frac{1}{\sum_{j=1}^{c}\left(D_{A_i^x}(x, v_i^x)/D_{A_i^x}(x, v_i^x)\right)^{2/(m-1)}} \quad [5.119]$$

This expression computes the degree of fulfillment of one rule relative to the other rules, and the sum of the membership degrees of all of the rules equals one, as in fuzzy clustering. Because of this constraint we class this method as probabilistic.

On the other hand, the degree of fulfillment of a rule can also be computed independently of the remaining rules, by applying a second transformation of the distance into the membership degree:

$$\beta_i(x) = \frac{1}{1 + D_{A_i^x}(x, v_i^x)} \quad [5.120]$$

In both cases, the distance is computed in relation to the cluster centre, which implies that the obtained multi-dimensional membership functions [5.119] and [5.120] have the same position (centre) in the antecedent space. However, the shape of the membership functions may differ significantly (Babuška 1998).

5.6.5. *Estimating consequent parameters*

There are several methods of obtaining the consequent parameters. Based on the geometrical interpretation of the TS model, the consequent parameters can be directly computed from the cluster prototypical points and the smallest eigenvectors of the cluster covariance matrices. This method assumes that errors are present in both the regressors and the regressand, and corresponds to the total least-squares solution of the local linearization around the cluster centre (Van Huffel and Vandewalle 1991).

A set of optimal parameters with respect to the model output can also be estimated from the identification data set by using the ordinary least-squares methods or the approach presented in section 5.3.4. This approach can be formulated as minimization of the total prediction error using the TS defuzzification formula [5.82], or as minimization of the prediction errors of the individual local models, solved as a set of c independent, weighted least-squares problems.

An example of identification of the consequent TS parameters by exploiting the total least-squares algorithm is shown in the following. This approach is usually preferred when the TS model serves as a predictor (Babuška 1998).

The consequent parameters a_i and b_i of the affine TS model from equation [5.116] can be derived from the geometrical structure of the clusters.

It is assumed that the collection of c clusters approximates the regression surface. These clusters can be approximately regarded as p-dimensional linear subspaces of the regression space. The eigenvector ϕ_{in} corresponding to the smallest eigenvalue λ_{in}, determines the normal vector to the hyperplane spanned by the remaining eigenvectors of that cluster.

The smallest eigenvector of the ith cluster will be denoted ϕ_i, omitting the subscript n. Recall that $z^T = [x^T, y]$ is the data vector and v_i is the cluster's prototype. The implicit normal form of the consequent's hyperplane is given by:

$$\phi_i (z - v_i) = 0 \tag{5.121}$$

This expression states that the product of the normal vector ϕ_i with any vector belonging to the hyperplane equals zero. For the following discussion, it is convenient to partition the prototype v_i into a vector v_i^x corresponding to the regressor x, and a scalar v_i^y corresponding to the regressand y: $v_i^T = \left[(v_i^x)^T, v_i^y \right]$. The smallest eigenvector is partitioned in the same way as the cluster centre, that is : $\phi_i^T = \left[(\phi_i^x)^T, \phi_i^y \right]$. Equation [5.121] can now be written as:

$$\left[(\phi_i^x)^T, \phi_i^y \right] \left(\left[x^T, y \right]^T - \left[(v_i^x)^T, v_i^y \right]^T \right) = 0 \tag{5.122}$$

Carrying out the inner product leads to the following equality:

$$(\phi_i^x)^T (x - v_i^x) + \phi_i^y (y - v_i^y) \tag{5.123}$$

from which, by a simple algebraic manipulation, an explicit equation for the hyperplane is obtained:

$$y = \frac{-1}{\phi_i^y} (\phi_i^x)^T x + \frac{1}{\phi_i^y} \phi_i^T v_i \tag{5.124}$$

By comparing the above expression with the affine consequent of the TS rule from equation [5.116], the expressions for a_i and b_i directly follow:

$$a_i = \frac{-1}{\phi_i^y} \phi_i^x, \tag{5.125}$$

$$b_i = \frac{1}{\phi_i^y} \phi_i^T v_i. \tag{5.126}$$

Although these equations have been derived from the geometrical interpretation of the clusters, it can be shown that a_i, given by equation [5.125] is obtained as a solution of a weighted Total Least Squares (TLS) problem, defined locally around the cluster centre v_i. The weights are the membership degrees contained in the ith row of the fuzzy partition matrix. Hence v_i is seen as a local operating point for the model. To obtain the affine linear form used in the TS rules [5.116], the offset parameters b_i are calculated using the estimates a_i and the cluster centre v_i.

After the clustering of the data has been obtained, data subsets can be processed according the Frisch scheme identification procedure (Frisch 1934; Beghelli *et al.* 1990), in order to estimate the TS parameters for each affine submodels, according to the rules presented in section 5.4.5 (Simani *et al.* 1998, 1999; Fantuzzi *et al.* 1998; Rovatti *et al.* 2000; Simani 2000).

The system identification with noisy measurements presented in section 5.3.2 can be applied and slightly adapted to handle affine models.

In order to identify the structure of the TS SISO model from equation [5.82] in the ith cluster with $i = 1, \cdots, c$ and c clusters, the following matrices can be defined:

$$X_k^{(i)} = \begin{bmatrix} y(k) & x_k^T(0) & 1 \\ y(k+1) & x_k^T(1) & 1 \\ \vdots & \vdots & \\ y(k+N_i-1) & x_k^T(N_i-1) & 1 \end{bmatrix} \qquad [5.127]$$

and therefore:

$$\Sigma_k^{(i)} = \left(X_k^{(i)}\right)^T X_k^{(i)}. \qquad [5.128]$$

In order to solve the noise-rejection problem in a mathematical framework, it is necessary to follow the assumptions (Frisch 1934; Kalman 1982b; Beghelli *et al.* 1990) that the noises $\tilde{u}(t)$ and $\tilde{y}(t)$ are additive on the input-output data $u^*(t)$ and $y^*(t)$ and region independent.

Under these assumptions, a positive-definite matrix $\Sigma_k^{(i)}$ associated to the sequences belonging with the ith cluster can be expressed as the sum of two terms $\Sigma_k^{(i)} = \Sigma_k^{*(i)} + \bar{\tilde{\Sigma}}_k$ where:

$$\bar{\tilde{\Sigma}}_k = \text{diag}[\bar{\tilde{\sigma}}_y I_{k+1}, \bar{\tilde{\sigma}}_u I_k, 0] \geq 0 \qquad [5.129]$$

The solution of the above identification problem requires the computation of the unknown noise covariance values $\bar{\tilde{\sigma}}_u$ and $\bar{\tilde{\sigma}}_y$, which can be achieved by solving the following relation:

$$\Sigma_k^{*(i)} = \Sigma_k^{(i)} - \tilde{\Sigma}_k \geq 0 \qquad [5.130]$$

in the variables $\tilde{\sigma}_u, \tilde{\sigma}_y$, where $\tilde{\Sigma}_k = \text{diag}[\tilde{\sigma}_y I_{k+1}, \tilde{\sigma}_u I_k, 0]$.

It is worth noting that all of the surfaces of the type defined by equation [5.130] have at least one common point, that is, point $(\bar{\tilde{\sigma}}_u, \bar{\tilde{\sigma}}_y)$ corresponding to the true variances of the noise affecting the input and output data.

The search for a solution for the identification problem can therefore start from the determination of this point in the noise space, if the noise characteristics are common to all of the clusters and all of the assumptions regarding the Frisch scheme are satisfied (independence between input-output sequences, additive noise, noise whiteness).

In real cases, these assumptions have to be relaxed, thus no common point can be determined among surfaces $\Gamma_n^{(i)} = 0$ in the noise plane, and a unique solution to the identification problem cannot be obtained.

In this situation, the local fuzzy model identification can be performed by finding the point $(\tilde{\sigma}_u, \tilde{\sigma}_y) \in \Gamma_{n+1}^{(i)} = 0$ that makes $\Sigma_{n+1}^{*(i)}$ closer to the double singular condition. It leads to the determination of the common point of the surfaces when the assumptions of the Frisch scheme are not violated.

Moreover, for each cluster, different noises $(\bar{\tilde{\sigma}}_u^{(i)}, \bar{\tilde{\sigma}}_y^{(i)})$ and the following relation should be rewritten as:

$$\Sigma_n^{*(i)} = \Sigma_n^{(i)} - \tilde{\Sigma}_n^{(i)} \geq 0 \qquad [5.131]$$

where $\tilde{\Sigma}_n^{(i)} = \text{diag}[\bar{\tilde{\sigma}}_u^{(i)} I_{n+1}, \bar{\tilde{\sigma}}_y^{(i)} I_n, 0]$ while $(\bar{\tilde{\sigma}}_u^{(i)}, \bar{\tilde{\sigma}}_y^{(i)})$ represent the variances of input and output additive noises in the ith cluster.

Finally, the matrices $\tilde{\Sigma}_n^{(i)}$ can therefore be built and the parameter of the model in each cluster determined by means of relation:

$$(\Sigma_n^{(i)} - \tilde{\Sigma}_n^{(i)})a^{(i)} = 0 \quad \text{for } i = 1, \ldots, c \qquad [5.132]$$

for a number of c clusters. This completes the multiple-model identification procedure in the fuzzy environment.

5.7. Conclusion

Several off-line identification methods have been presented in this chapter for the estimation of both linear and nonlinear static and dynamic systems from data affected by noise. Linear, piecewise affine (linear) and fuzzy models that can be used for fault diagnosis purposes were discussed.

For the case of piecewise affine and fuzzy models, the multiple-model approach consists of using several local affine submodels, each describing a different operating condition of the process. The identification algorithm exploited to estimate the parameters and orders of the local affine submodels is based on the well-established Frisch Scheme method for linear systems. For the nonlinear case, in order to obtain a continuous piecewise affine prototype, describing the input-output behavior of the process, continuity constraints between local linear dynamic models have to be forced. For non-fuzzy models, such a continuity constrained problem was solved by using an optimization technique. The properties of the solutions obtained by the Frisch Scheme enhance the fulfillment of the constraints.

On the other hand, in order to identify a nonlinear process, neural networks, fuzzy and neuro-fuzzy modeling and identification methods can be exploited. By means of product space fuzzy clustering, a data set generated by the nonlinear system can be partitioned into fuzzy subsets of data that are locally described by linear sub models. Prior to clustering, the regression structure of the model must be selected, in order to properly represent the system dynamics. After the structure is determined, clustering in the product space of the regressors and of the regressand can be applied to partition the data. Since each cluster serves as a local linear model of the system, the clustering algorithm must be capable of detecting clusters which lie in linear subspaces.

Therefore, for the final construction of the Takagi–Sugeno fuzzy models, two main tasks can be distinguished: the generation of the antecedent membership functions, and the estimation of the consequent parameters. The consequent parameters can be computed from the cluster centers and covariance matrices, or can be estimated by least-squares methods, as well as by local application of the Frisch scheme identification procedure for linear models.

Finally, fuzzy identification algorithms exploited to both generate fuzzy partitions and estimate the Takagi–Sugeno structure were achieved by using proper fuzzy clustering and identification procedures.

5.8. References

Akaike, H. (1974). A new look at the statistical model identification. *IEEE Trans. Automatic Control*, 19(6), 716–723.

Babu, G. and Murty, M. (1994). Clustering with evolution strategies. *Pattern Recognition*, 27(2), 321–329.

Babuška, R. (1998). *Fuzzy Modeling for Control.* Kluwer Academic Publishers, Norwell, MA.

Babuška, R. (2000). Fuzzy Modelling and Identification Toolbox, version 3.1. Control Engineering Laboratory, Faculty of Information Technology and Systems, Delft University of Technology, Delft, The Netherlands [Online]. Available at: http://lcewww.et.tudelft.nl/b̃abuska.

Babuška, R. and Verbruggen, H.B. (1995). Identification of composite linear models via fuzzy clustering. In *Proceedings of the 3rd ECC'95*, Rome, Italy, 1207–1212.

Babuška, R., Keizer, R., Verhaegen, M. (1997). Identification of nonlinear dynamic systems as a composition of local linear parametric or state-space models. In *Proceedings of the SYSID'97*, Fukuoka, Japan, 675–680.

Backer, E. (1995). *Computer–Assisted Reasoning in Cluster Analysis.* Prentice Hall, New York.

Banks, S. and Kathur, S. (1989). Structure and control of piecewise linear system. *Int. J. of Control*, 50, 346–358.

Beghelli, S. and Soverini, U. (1992). Identification of linear relations from noisy data: Geometrical aspects. *System and Control Letters*, 18(5), 339–346.

Beghelli, S., Guidorzi, R.P., Soverini, U. (1990). The Frisch scheme in dynamic system identification. *Automatica*, 26(1), 171–176.

Beghelli, S., Castaldi, P., Guidorzi, R.P., Soverini, U. (1994a). A comparison between different model selection criteria in Frisch scheme identification. *Systems Science Journal*, 20(1), 77–84.

Beghelli, S., Castaldi, P., Soverini, U. (1994b). Dynamic Frisch scheme identification: time and frequency domain approaches. In *IFAC'94, 10th IFAC Symposium on System Identification.*

Beghelli, S., Guidorzi, R.P., Soverini, U. (1997). A frequencial approach to the dynamic Frisch scheme identification. In *ECC'97, 4th European Control Conference*, Brussels, Belgium.

Bemporad, A. and Morari, M. (1999). Control of systems integrating logic, dynamics, and constraints. *Automatica*, 35(3), 407–428.

Bezdek, J. (1980). A convergence theorem for the fuzzy isodata clustering algorithms. *IEEE Trans. Pattern Anal. Machine Intell.*, PAMI, 2(1), 1–8.

Bezdek, J. (1981). *Pattern Recognition with Fuzzy Objective Function.* Plenum Press, New York.

Bezdek, J. and Pal, S. (1992). *Fuzzy Models for Pattern Recognition: Methods that Search for Patterns in Data.* IEEE Press, New York.

Bezdek, J., Hathaway, R., Howard, R., Wilson, C., Windham, M. (1987). Local convergence analysis of a grouped variable version of coordinate descendent. *Journal of Optimization Theory and Applications*, 54(3), 471–477.

Billings, S.A. and Voon, W. (1983a). Structure detection and model validity test in the identification of nonlinear systems. *IEE Proceedings*, 130D(4), 193–199.

Billings, S.A. and Voon, W. (1983b). Structure detection and model validity tests in the identification of nonlinear systems. *IEE Proceedings*, 130(4), 193–200.

Billings, S.A. and Voon, W. (1986). Correlation based model validity test for nonlinear models. *International Journal of Control*, 44(1), 235–244.

Brown, M. and Harris, C. (1994). *Neurofuzzy Adaptive Modeling and Control.* Prentice Hall, Hemel Hempstead, UK.

Chen, S. and Billings, S.A. (1989). Representation of nonlinear systems: The NARMAX model. *Int. J. Control*, 49, 1013–1032.

Chen, S., Billings, S.A., Cowan, C.F., Grant, P. (1990). Practical identification of NARMAX model using radial basis functions. *Int. J. Control*, 52, 1327–1350.

Chen, S., Cowan, C., Grant, P. (1991). Orthogonal least squares learning algorithm for radial basis function networks. *IEEE Trans. Neural Networks*, 2(2), 302–309.

Cottle, R. (1982). Minimal triangulation of the 4-cube. *Discrete Mathematics*, 40, 25–29.

de Boor, C. (1978). *A Practical Guide to Splines*. Springer-Verlag, New York.

Delmaire, G., Cassar, P., Staroswiecki, M., Christophe, C. (1999). Comparison of multivariable identification and parity space techniques for FDI purpose in M.I.M.O. systems. In *ECC'99*, Karlsruhe, Germany.

DeSarbo, W. (1982). Gennclus: New models for general nonhierarchical clustering analysis. *Psychometrika*, 47(4), 449–476.

Duda, R. and Hart, P. (1973). *Pattern Classification and Scene Analysis*. John Wiley & Sons, New York.

Dunn, J. (1974). A fuzzy relative of the ISODATA process and its use in detecting compact well-separated clusters. *International Journal of Cybernetics and Systems*, 3(3), 32–57.

Fantuzzi, C. and Rovatti, R. (1996). On the approximation capabilities of the homogeneous Takagi–Sugeno model. *Proceedings of the 5th IEEE International Conference on Fuzzy Systems*, 1067–1072.

Fantuzzi, C., Rovatti, R., Simani, S., Beghelli, S. (1998). Fuzzy modeling with noisy data. In *EUFIT'98, vol. 3, The 6th European Congress on Intelligent Techniques and Soft Computing*, Aachen, Germany, 1615–1619.

Fantuzzi, C., Simani, S., Beghelli, S., Rovatti, R. (2002). Identification of piecewise affine models in noisy environment. *International Journal of Control*, accepted.

Friedmann, J. (1991). Multivariable adaptive regression splines. *The Annal of Statistics*, 1–141.

Frisch, R. (1934). *Statistical Confluenece Analysis by Means of Complete Regression Systems*, 5th edition, Economic Institute, University of Oslo.

Guidorzi, R.P. (1975). Canonical structures in the identification. *Automatica*, 11, 361–374.

Guidorzi, R.P. (1981). Invariants and canonical forms for system structural and parametric identification. *Automatica*, (17), 117–133.

Guidorzi, R.P., Losito, P., Muratori, T. (1982). The range error test in the structural identification of linear multivariable systems. *IEEE Transactions on Automatic Control*, 27(5), 1044–1054.

Gustafson, D.E. and Kessel, W.C. (1979). Fuzzy clustering with a variance covariance matrix. In *Proceedings IEEE CDC'79*, San Diego, CA, 161–166.

Hadamard, J. (1964). *La théorie des équations aux dérivées partielles*. Editions Scientifiques, Pekin.

Haiman, M. (1991), A simple and relatively efficient triangulation of the n-cube. *Discrete Computational Geometry*, 6, 287–289.

Hathaway, R. and Bezdek, J. (1983). Switching regression models and fuzzy clustering. *IEEE Trans. of Fuzzy Systems*, 1(3).

Hunt, J., Lee, M., Price, C. (1992a). An introduction to qualitative model-based reasoning. In *Preprints IFAC/IFIP/IMACS Int. Sym. on Artificial Intelligence in Real-Time Control*, Delft, The Netherlands.

Hunt, K., Sbarbaro, D., Zbikowki, R., Gawthrop, P. (1992b). Neural networks for control systems: A survey. *IEEE Trans. Neural Networks*, 28, 1083–1112.

Jager, R. (1995). Fuzzy logic in control. PhD Thesis, Delft University of Technology, Delft, The Netherlands.

Jain, A. and Dubes, R. (1988). *Algorithms for Clustering Data*. Prentice Hall, Englewood Cliffs, NJ.

Jazwinski, A.H. (1970). *Stochastic Processes and Filtering Theory*. Academic Press, New York.

Johansen, T.A. (1996). Robust identification of Takagi–Sugeno–Kang Fuzzy models using regularization. In *5th IEEE International Conference on Fuzzy Systems*, New Orleans, LA.

Johansen, T.A. and Foss, B. (1993). Constructing NARMAX models using ARMAX models. *Int. J. Control*, 58(5), 1125–1153.

Juditsky, A., Hjalmarsson, H., Beneviste, A., Delyon, B., Ljung, L., Sjöberg, J., Zhang, Q. (1995). Nonlinear black-box modeling in system identification: A mathematical foundation. *Automatica*, 31(12), 1691–1724.

Kalman, R.E. (1982a). Identification from real data. In *Current Developments in the Interface: Economics, Econometrics, Mathematics*, Hazewinkel, M., Rinnoy Kan, A.H.G. (eds). D. Reidel, Dordrecht, The Netherlands.

Kalman, R.E. (1982b). System identification from noisy data. In *Dynamical System II*, Bednarek, A.R., Cesari, L. (eds). Academic Press, New York.

Kalman, R.E. (1984). Identification of noisy systems. In *50th Anniversary Symp. Steklov Institute of Mathematics*, U.S.S.R. Academy of Sciences, Moscow.

Kalman, R.E. (1990). Nine lectures on identification. *Lecture Notes on Economics and Mathematical System*, Springer-Verlag, Berlin.

Klir, G.J. and Yuan, B. (1995). *Fuzzy Sets and Fuzzy Logic: Theory and Applications*. Prentice Hall, Upper Saddle River, NJ.

Kosko, B. (1994). Fuzzy systems as universal approximators. *IEEE Transactions on Computers*, 43, 1329–1333.

Lee, Y., Hwang, C., Shih, Y. (1994). A combined approach to fuzzy model identification. *IEEE Trans. Sys. Man, Cybern.*, 24(5), 736–744.

Leontaritis, I. and Billings, S.A. (1985a). Input-output parametric models for nonlinear systems part II: Stochastic nonlinear systems. *Int. J. Control*, 41(2), 329–344.

Leontaritis, I. and Billings, S.A. (1985b). Input-output parametric models for nonlinear systems part I: Deterministic nonlinear systems. *Int. J. Control*, 41(2), 303–328.

Ljung, L. (1999). *System Identification: Theory for the User*, 2nd edition. Prentice Hall, Englewood Cliffs, NJ.

Luenberger, D.G. (1971). An introduction to observers. *IEEE Transactions on Automatic Control*, AC-16(6), 596–602.

Mara, P. (1976). Triangulation for the cube. *Journal of Combinatorial Theory (A)*, 20, 170–177.

McGraw, K. and Harbisson-Briggs, K. (1989). *Knowledge Acquisition: Principles and Guidelines*. Prentice Hall, Englewood Cliffs, NJ.

Morozov, V. (1984). *Methods for Solving Incorrectly Posed Problems*. Springer, Berlin.

Murray-Smith, R. and Johansen, T.A. (1997). *Multiple Model Approaches to Nonlinear Modeling and Control*. Taylor & Frencis, London.

Norton, J. (1986). *An Introduction to Identification*. Academic Press, London.

Pettit, N. and Wellstead, P. (1995). Analyzing piecewise linear dynamical systems. *IEEE Control System*, 43–50.

Priestly, M. (1988). *Non–linear Non–stationary Time Series Analysis*. Academic Press, London.

Rissanen, J. (1978). Modeling by shortest data description. *Automatica*, (14), 465–471.

Rovatti, R. (1996). Takagi–Sugeno models as approximators in Sobolev norms: The SISO case. In *5th IEEE International Conference on Fuzzy Systems*, New Orleans, LA.

Rovatti, R., Borgatti, M., Guerrieri, R. (1998). A geometric approach to maximumspeed n-dimensional linear interpolation in rectangular grids. *IEEE Transactions on Computers*, 47, 894–899.

Rovatti, R., Fantuzzi, C., Simani, S. (2000). High-speed DSP-based implementation of piecewise-affine and piecewise-quadratic fuzzy systems. *The Signal Processing Journal*, 80(6), 951–963.

Sallee, J. (1982). A triangulation of the n-cube. *Discrete Mathematics*, 40, 81–86.

Sallee, J. (1984). Middle-cut triangulations of the n-cube. *SIAM Journal on Algebraic and Discrete Methods*, 5, 407–419.

Seber, G. and Wild, C. (1989). *Nonlinear Regression*. John Wiley & Sons, New York.

Simani, S. (2000). Multi model based fault diagnosis of a sugar cane crushing process. In *SAFEPROCESS2000, vol. 2, 4th Symposium on Fault Detection Supervision and Safety for Technical Processes*, Budapest, Hungary, 657–662.

Simani, S., Fantuzzi, C., Rovatti, R., Beghelli, S. (1998). Noise rejection in parameters identification for piecewise linear fuzzy models. In *WCCI'98, FUZZ-IEEE'98, 1998 IEEE International Conference on Fuzzy Systems*, Ancorage, Alaska, 378–383.

Simani, S., Fantuzzi, C., Rovatti, R., Beghelli, S. (1999). Parameter identification for piecewise linear fuzzy models in noisy environment. *International Journal of Approximate Reasoning*, 1–2(22), 149–167.

Sjöberg, J., Zhang, Q., Ljung, L., Beneviste, A., Delyon, B., Glorennec, P.-Y., Hjalmarsson, H., Juditsky, A. (1995). Nonlinear black-box modeling in system identification: A unified overview. *Automatica*, 31(12), 1691–1724.

Skeppstedt, A., Ljung, L., Millnert, M. (1992). Construction of composite models from observed data. *Int. J. Control*, 55, 141–152.

Söderström, T. and Stoica, P. (1987). *System Identification*. Prentice Hall, Englewood Cliffs, NJ.

Sontag, E. (1981). Nonlinear regulation: The piecewise linear approach. *IEEE Trans. on Automatic Control*, 26, 346–358.

Strang, G. and Fix, G. (1973). *An Analysis of the Finite Element Method*. Prentice-Hall, Englewood Cliffs, NJ.

Sugeno, M. and Kang, G. (1988). Structure identification of fuzzy model. *Fuzzy Set and Systems*, 28, 15–33.

Takagi, T. and Sugeno, M. (1985). Fuzzy identification of systems and its application to modeling and control. *IEEE Transaction on System, Man and Cybernetics*, 15(1), 116–132.

Tikhonov, A. and Arsenin, V. (1977). *Solution of Ill-posed Problems*. Winston and Wiley, Washington.

Van Huffel, S. and Vandewalle, J. (1991). *The Total Least Squares Problem: Computational Aspects and Analysis*. SIAM, Philadelphia, PA.

Verhaegen, M. and Dewilde, P. (1992). Subspace model identification. Part 1: The output error state-space model identification class of algorithms. *International Journal of Control*, 56(1), 1187–1210.

Wang, L.-X. (1992). Fuzzy systems are universal approximators. In *Proc. First IEEE Int. Conf. on Fuzzy Syst.*, San Diego, CA.

Wang, L.-X. (1995). The design and analysis of fuzzy identifiers of nonlinear dynamic systems. *IEEE Transactions on Automatic Control*, 40(1), 11–23.

Zeng, X.-J. and Singh, M. (1996). Approximation accuracy analysis of fuzzy systems as function approximators. *IEEE Transactions on Fuzzy Systems*, 4, 44–63.

6

The Artificial Intelligence Approach to Model-based Diagnosis

Belarmino PULIDO, Carlos J. ALONSO-GONZÁLEZ, Anibal BREGON

Department of Computer Science, University of Valladolid, Spain

6.1. Introduction

The Artificial Intelligence (AI) discipline has always devoted a lot of attention to the high-level cognitive task of diagnosis. From the very beginning, diagnosis has been considered a classification problem that links symptoms with causes. Early works with expert, rule-based systems in the field of medicine – such as MYCIN (Shortliffe 1976) – followed this reasoning path, either for medical diagnosis or for identifying man-made system malfunctions.

The definition provided by Console places emphasis on the cognitive task issue, and its relation to other high-level tasks such as repair and reconfiguration:

> *Diagnosis: Diagnosis is the task, that given a system and a set of observations from a faulty situation, is able to determine what is wrong in the system (and what would be necessary to repair in order to bring back the system to a normal state)* (Console 2000).

With the AI approach to building intelligent systems for diagnosis, there is a general trend to clearly isolate: the source of knowledge, to identify root causes and

Diagnosis and Fault-tolerant Control 1,
coordinated by Vicenç PUIG and Silvio SIMANI.

symptoms; the process, to generate the symptoms from the set of available observations; and, finally, the reasoning process, to link symptoms and causes (Balakrishnan and Honavar 1998). This is a rather different approach to the classical characterization proposed by Venkatasubramanian *et al.* (2003) and Gertler (1998), commonly referenced in the Control Theory literature.

In this classification, there are four major categories:

– *Knowledge-based systems (KBS)* (Jackson 1998; Stefik 2014): these use available knowledge about the domain, mainly expertise-based, avoiding the use of explicit mathematical models. The knowledge provides mostly causal relations between symptoms and causes, which are explored using forward or backward chaining algorithms, from symptoms to causes or vice versa. The main drawback is that every solution is domain specific and it is unable to learn from new experiences, but for many years KBS have been the main source of diagnosis systems in many domains: medical (MYCIN, Shortliffe 1976), electronic system troubleshooting (SOPHIE, Brown *et al.* 1982), oil extraction (Kahn and McDermott 1986), financial (Matsatsinis *et al.* 1997), etc.

– *Case-based reasoning (CBR) systems* (Kolodner 2014; Watson 1998): implemented as a collection of cases that merely state the symptoms and the causes from previous situations with no explicit model for inference. The solutions are provided in a four "R" cycle: "Retrieve" similar cases (using some kind of metric) from the case-base; "Reuse" previous solutions; "Revise" or adapt the solution depending on the true results; and "Retain" the new case if it is new or interesting enough. The approach provides a simple, lazy learning approach for new situations, but requires a lot of cases to provide a capable solution.

– *Machine learning systems* (Mitchell 2006; Witten *et al.* 2016): these use previous examples (given as observations and previous diagnoses) to extract new relations between observations (symptoms) and diagnoses. The main drawback is that the induced models lack self-explanation capabilities, but also that a lot of previous samples are required to produce a domain-specific solution.

– *Model-based systems* (Reiter 1987; Hamscher *et al.* 1992): these rely upon existing analytic models (based on the law of physics) about the structure and behavior of the system to be diagnosed (Davis 1982). The system models are used to estimate the desired, fault-free behavior of the system given existing observations. Any deviation in these predictions from current observed behavior is considered a symptom and the components used to perform the estimation are considered to be potential fault candidates. The main advantage of the approach is the ability to isolate the models from the reasoning process, thus providing straightforward model reusability. The main drawback is that real systems need complex models to provide accurate diagnoses.

As is evident, there are many options that can be used depending on the amount of available knowledge. Data-driven models derived from existing observations were analyzed in Chapter 5, and CBR can be seen as a specific case of "lazy" learning or learning from instances. Consequently, in this chapter we focus on the two most classical AI approaches to diagnosis: KBS and model-based diagnosis. The first approach is no longer a hot topic of research in the field but has provided a complete set of successful diagnosis systems. The second approach is the most sound from a theoretical point of view, providing highly generalizable solutions, but has provided only very specific tools for diagnosing rather specific domains. Later in the chapter, we provide the logical and computational foundations of early works in the DX approach, together with the early extensions to include fault modes in the models, leading to more precise identification.

6.2. Case studies

In this chapter, we consider two different case studies: (i) a static system (full adder) to illustrate the fundamental concepts of consistency-based diagnosis; and (ii) a continuous system (a three-tank system) to illustrate the extensions of consistency-based diagnosis for dynamic systems.

Regarding the static system, we use a binary full adder that is a classical example in the CBD literature taken from Reiter (1987). The full adder, shown in Figure 6.1 is made up of two AND gates, two XOR gates and one OR gate, that we name A1, A2, X1, X2 and O1, respectively. All five components might be faulty. The following set of equations describes the nominal operation of the full adder:

$$
\begin{aligned}
&\text{X1: } x = \text{XOR}(a, b),\\
&\text{X2: } s = \text{XOR}(x, c_{in}),\\
&\text{A1: } y = \text{AND}(c_{in}, x),\\
&\text{A2: } z = \text{AND}(a, b),\\
&\text{O1: } c_{out} = \text{OR}(y, z).
\end{aligned}
$$

The system has five observable variables, a, b and c_{in} which are the inputs, and s and c_{out} which are the outputs. It also has three unobserved variables, x, y and z.

In terms of continuous systems, we will use a three-tank system originally presented in Ding (2014) and adapted in Escobet *et al.* (2019). Figure 6.2 shows the schema of the three-tank system.

The system is made up of three connected tanks, $\{T_1, T_2, T_3\}$, that have identical height and cross-sectional area. The system has only one input, q_i, that drains water into T_1. Two pipes, q_{12} and q_{23}, connect the three tanks and drain liquid from T_1 to T_2, and from T_2 to T_3, respectively. Finally, it has an output flow q_{30} from T_3. The system has three sensors $LT1$ and $LT2$ which measure the level in tanks T_1 and T_3, respectively; and $FT1$ which measures the flow through pipe q_{23}. A total of six possible faults have been considered for the case study, three pipe blockages and three-tank leakages, which have been modeled using the parameters $Stuck_{q_{jk}}$ and $Leakage_{T_j}$. Modeling the system faults in this way means that we can deal with both multiplicative and additive faults in the model, just by setting the corresponding parameters to 0 or 1.

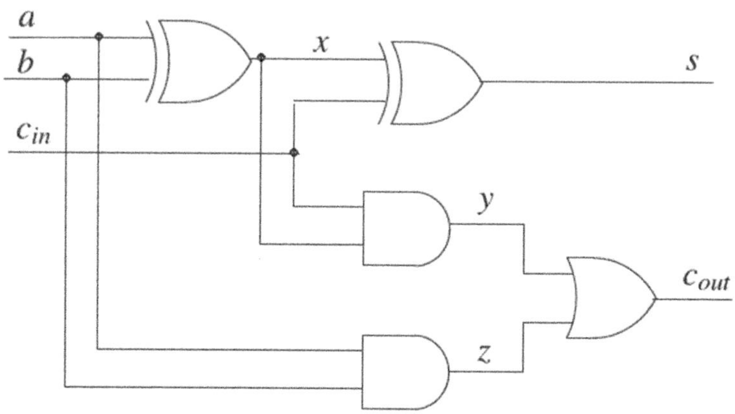

Figure 6.1. *The full adder (source: Reiter 1987)*

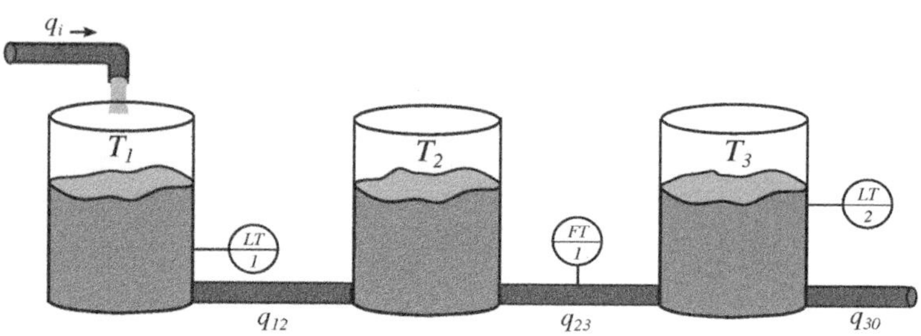

Figure 6.2. *The three-tank system (source: Escobet* et al. *2019)*

The following systems model (obtained from Escobet *et al.* 2019) describes the behavior of the system. The change in the level in each tank, $\dot{h}_{T_j}(t)$ for $j = 1, \cdots, 3$, is computed according to mass balances:

$$e1 : \dot{h}_{T_1}(t) = \frac{q_i(t) - q_{12}(t) - q_{f_1}(t)}{A},$$

$$e2 : \dot{h}_{T_2}(t) = \frac{q_{12}(t) - q_{23}(t) - q_{f_2}(t)}{A},$$

$$e3 : \dot{h}_{T_3}(t) = \frac{q_{23}(t) - q_{30}(t) - q_{f_3}(t)}{A}$$

with A being the cross-sectional area of the tanks. The flow between tanks, $\{q_{12}(t), q_{23}(t)\}$, the output flow, $\{q_{30}(t)\}$, and the leakage flows, $\{q_{f_1}(t), q_{f_2}(t), q_{f_3}(t)\}$, are modeled as:

$$e4 : q_{12}(t) = (1 - Stuck_{q_{12}})S_{p_1}\text{sign}(h_{T_1}(t) - h_{T_2}(t))\sqrt{2g \mid h_{T_1}(t) - h_{T_2}(t) \mid},$$

$$e5 : q_{f_1}(t) = Leakage_{T_1}\sqrt{2gh_{T_1}(t)},$$

$$e6 : q_{23}(t) = (1 - Stuck_{q_{23}})S_{p_2}\text{sign}(h_{T_2}(t) - h_{T_3}(t))\sqrt{2g \mid h_{T_2}(t) - h_{T_3}(t) \mid},$$

$$e7 : q_{f_2}(t) = Leakage_{T_2}\sqrt{2gh_{T_2}(t)},$$

$$e8 : q_{30}(t) = (1 - Stuck_{q_{30}})S_{p_3}\sqrt{2gh_{T_3}(t)},$$

$$e9 : q_{f_3}(t) = Leakage_{T_3}\sqrt{2gh_{T_3}(t)}$$

with S_{p_i} being the cross-sectional area of the pipes. It is assumed that the measured variables are h_{T_1}, h_{T_3} and q_{23} and the input flow q_i is known. It is also assumed that the initial water level in the three tanks is zero, but this can be configured for each diagnosis scenario.

In the following, the variable t in parentheses will be omitted, so that no confusion can arise.

6.3. Knowledge-based diagnosis systems

A KBS is a computer system that handles an explicit and symbolic representation of the knowledge necessary to solve a problem. Furthermore, the behavior of the system must be generated from, and be coherent with, the knowledge thus represented. From a structural point of view, a KBS has at least two components: the knowledge base and the inference engine. The knowledge base contains a declarative representation of the knowledge the system has. The inference engine is responsible for using that knowledge to solve the problem, usually making inferences using the knowledge and current observations.

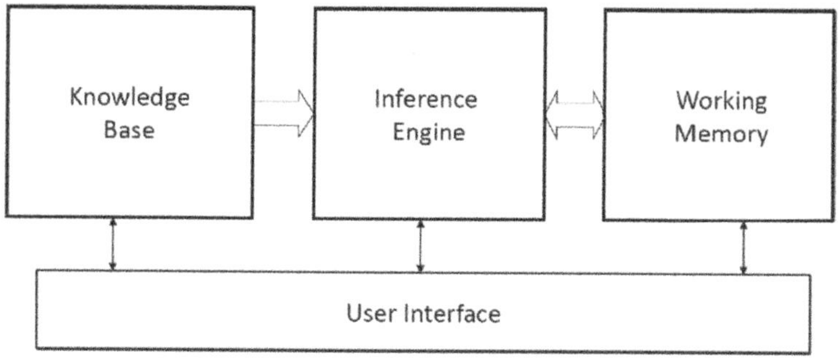

Figure 6.3. *Simple KBS architecture*

A working KBS may have additional modules, for instance for knowledge acquisition or generation of an explanation, but at least two other modules are usually found: the working (or dynamic) memory and some kind of user interface. Figure 6.3 shows the basic architecture of a KBS and the information flow between its modules. From the end user point of view, the working memory contains initial facts and any additional facts that the inference engine is able to deduce using the content of the knowledge base.

The most popular knowledge representation language is production rules. Production rules have a very simple syntax that can be stated as:

If<*antecedent*>**Then**<*consequent*>

Semantic rules are also simple: if the <antecedent> is true – according to current working memory content – then the <consequent> is also true. The inference engine exploits the syntax of the rule to perform inferences, concluding the <consequent>, which procedurally translates into adding the consequent content to the working memory.

Figure 6.4 shows a MYCIN rule form (Buchanan and Shortliffe 1984). MYCIN is a system designed to diagnose and propose a therapy for a very specific bacterial infection: meningitis. It was a historical landmark in expert systems development and became the paradigm of rule-based expert systems. MYCIN coded its diagnosis knowledge using production rules. It also introduced the Certainty Factors model as a simple formalism to manage uncertainty – the value 0.7 shown in the consequent of this rule.

The rule in Figure 6.4 allows the inference engine to conclude that "the identity of the organism is streptococcus" if and only if all of the premises of the antecedent are in the working memory.

If

(1) the stain of the organism is gram-positive
(2) the morphology of the organisms is coccus
(3) the growth configuration of the organism is chains

Then

there is a suggestive evidence (0.7)
that the identity of the organisms is streptococcus

Figure 6.4. *Example of a MYCIN production rule rewritten in the form of natural language*

The inference engine can organize the search for a solution in two different ways. The direct interpretation of the rules yields to the *forward chaining* search, where from the initial content of the working memory, the inference engine adds new information – usually called facts – to the working memory until a solution is found, or until no more rules can be applied. But the search can be performed in the opposite direction: from the initial goal to the initial facts. In this case, the inverse interpretation is exploited: in order for the <consequent> to be true, the <antecedent> must also be true. This goal-directed search is called *backward chaining*. Although the direction of the search does not theoretically change what the inference engine can derive, in practice backward chaining provides some additional control for the search process. Backward chaining is particularly useful when not all of the initial data are already available because the search process provides an implicit strategy for new data acquisition, which focuses around the goal that is currently being obtained. This is the main reason why the backward chaining search direction is performed by many diagnosis KBS.

6.3.1. *Diagnosis task and system model*

In its most general approach, the diagnostic task can be considered a process that, based on observations, generates hypotheses capable of explaining faulty behavior in order to select repair actions.

Any approach to the diagnostic task requires a system model that allows relationships to be established between data, hypotheses and repairs. The system model must also support the basic reasoning operations to solve the problem, which are essentially the generation, test and discrimination of hypotheses.

KBS are characterized by using explicit models, mostly qualitative in nature, to perform the diagnostic task. Furthermore, almost all of these models can be considered classification models, including causal networks. Classification models assume that the diagnostic task can be considered a classification problem. To do this, it is necessary to list the possible failures of a system, which are considered classes. In its simplest form, classes are characterized by symptoms, which are abstractions of the basic data. The solution to the problem is obtained when the classes compatible with the available observations have been identified.

These types of models are usually conceptually simple, as well as the solution methods that manipulate them. This explains its popularity: many of the first-generation expert systems that addressed diagnostic problems treated it as a classification problem. Systems such as MYCIN, INTERNIST or CENTAUR perform a classification task. All of these examples belong to the domain of medicine, in which first principles models or quantitative numerical models are usually lacking. However, there are also examples from other domains: MORE (extraction of hydrocarbons), MUD (conditions of drilling fluids), or SOPHIE (system pioneer in the detection of faults in electronic circuits).

A simple formal model for classification problems can be found in Reggia *et al.* (1983) and Peng and Reggia (1990). Stefik (2014) provides various classification models, along with associated troubleshooting methods. In this chapter we will illustrate these types of models with the historically important "Heuristic Classification" proposed by Clancey (1985), which explains many of the models and reasoning processes used by KB expert systems.

6.3.1.1. *Clancey "heuristic classification"*

Clancey's heuristic classification is a classification method that matches abstract data to abstract solutions. Hence, the problem-solving method consists of the three basic steps shown in Figure 6.5:

- **abstraction**: where semantically richer abstract data are built from raw data;
- **matching**: where abstract solution are selected considering current abstract data;
- **refinement**: where the abstract solutions are refined to specific solutions, possibly using more data.

Essentially, Clancey's heuristic classification model assumes that the knowledge needed to perform the classification can be acquired from experts in the problem and that it is heuristic in nature. Applied to a diagnostic problem, it associates symptoms with groups of failures, which are refined with additional observations. The system model is reduced to the knowledge necessary to abstract the symptoms of interest, the grouping of failures into families, if possible, supplied with a hierarchy that facilitates the search for specific solutions, and the association between symptoms and failure families.

Clancey proposes several types of data abstraction that are well illustrated by a fragment of the particularization of this model in the MYCIN expert system, as shown in Figure 6.6.

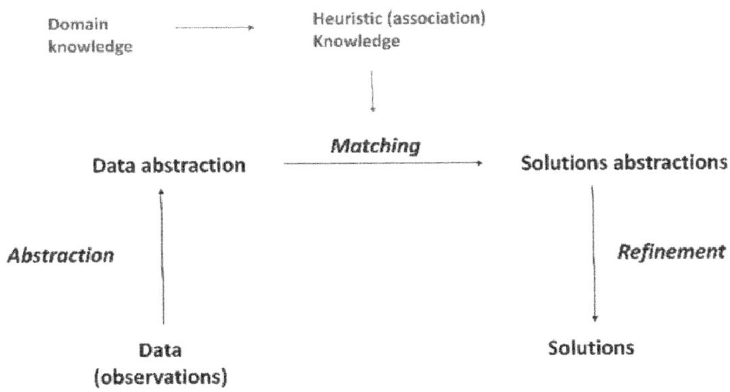

Figure 6.5. *Heuristic classification (from Clancey 1985)*

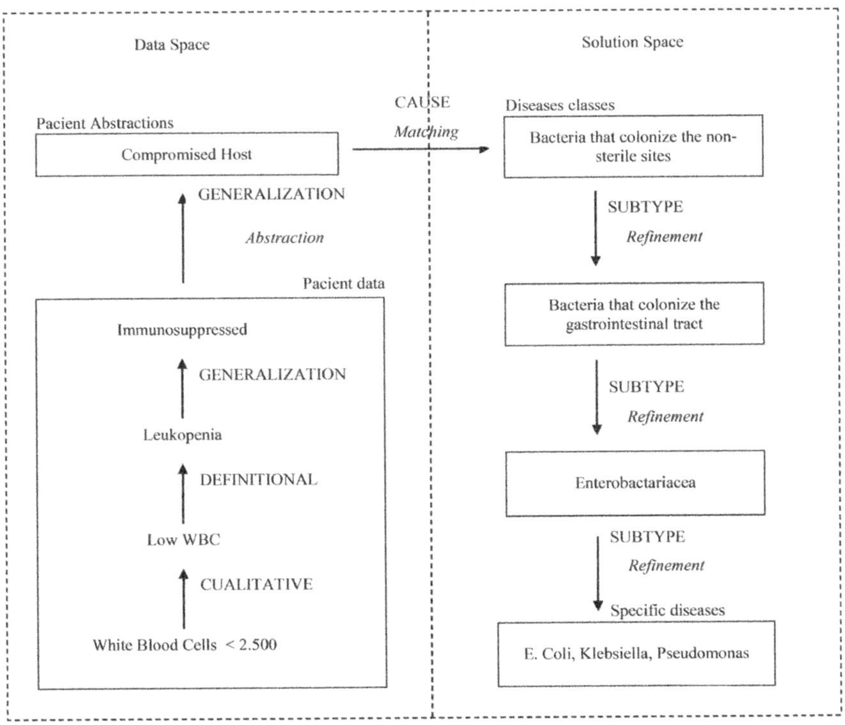

Figure 6.6. *Heuristic classification in MYCIN (adapted from Clancey 1985)*

Note, however, that all of the steps of the model shown in Figure 6.6 are not always made explicit in MYCIN rules. Figure 6.7 shows the rule that actually codifies the aforementioned model, which does not make the model explicit. It can be seen that the rule also codes strategic knowledge in the third and fourth premises, as well as knowledge of the context where the rule applies in the first and second premises of the antecedent.

If

(1) the infection is meningitis
(2) the type of infection is bacterial
(3) no organism has been directly identified
(4) complete blood count results are available
(5) the white blood count is less than 2.500

Then

there is suggestive evidence
the identity of the organisms is E. Coli (0.75),
Pseudomonas (0.5), Klebsiella (0.5)

Figure 6.7. *MYCIN production rule for the heuristic model of Figure 6.6*

6.3.2. *Diagnosis of physical devices*

The same classification models and problem-solving methods can be used for the diagnosis of physical devices, as long as there is a body of contrasted experience in their diagnoses.

However, some considerations have to be taken into account, especially regarding dynamic systems. KB diagnosis systems, mainly due to their dependence on experience, are usually designed for the desired operation point of the systems, which is usually a stationary state, frequently commanded by an automatic control system. The set of rules that apply to the nominal operation point may no longer be valid in a different operation point, especially when we are dealing with nonlinear systems, because nonlinearities might change the nature of the observable symptoms. Hence, if we want to develop a KBS diagnosis system capable of coping with a nonlinear system in different operation points, we will have to develop models for each different operation point and apply the model for the operation point at hand.

The raw signals from dynamical systems are generally abstracted for reasoning. It is the equivalent to Clancey's abstraction step. Usually, three abstract values are enough, sometimes five and very seldom seven or more. For three abstract values, we use the symbols $\{0, +, -\}$ to indicate that a magnitude is normal, higher than normal, or lower than normal. The same symbols can be applied to qualitatively describe the first derivative of a magnitude.

We will illustrate the approach on the second case study, with two simple rules for diagnosing a leak or a partial blockage of a pipe in the system. We will assume no sensor fault and single fault hypothesis.

For this system, we assume that the input, q_i, is a constant, whose value determines the stationary values of the measured magnitudes, $LT1$, $FT1$ and $LT2$. These values are needed to determine the abstract values of the measurements.

A simple rule can be used to detect an upstream leak or pipe blockage and to establish it as a first abstract solution:

If (1) $FT1 = -$

Then there is an upstream leak or blockage in the system.

This abstract solution can be refined using the information provided by other sensors. For instance:

If

(1) there is an upstream leak or blockage in the system

(2) $LT2 = 0$

Then there is a leak at $T2$.

6.3.3. *Limits of KBS for diagnosis of physical devices*

KBS offer a mature technology. Several Ontologies, like SAREF and its extensions (Daniele *et al.* 2018; Fernández-Izquierdo *et al.* n.d.) or ifcOWL (Pauwels and Terkaj n.d.), are available to facilitate the development of knowledge models. They are particularly suitable when there is experience in diagnosing a system and there are no other forms of knowledge useful for the task. As a result, they are not adequate for new devices or devices whose configuration or components might change.

Nevertheless, KBS face some problems that are related to their experience dependence. Knowledge acquisition is usually a demanding task that depends critically on the availability of experts. Even then, they suffer from a strong dependency on the device, since the experience is linked to each specific device. This issue, known as device dependence, is a severe limitation because different components with similar specifications can show different failure patterns.

KBS also suffer from the limits of the problem-solving approach: classification. Multiple faults are difficult to consider and unforeseen failures are out of their scope, contributing to the brittleness of the diagnosis system.

Finally, it must be considered that the maintenance of the consistency of large knowledge bases is critical and should be considered in the lifecycle of the final product.

6.4. Model-based diagnosis

Once the KBS used for diagnosis, troubleshooting or reconfiguration were unable to overcome their limitations (domain-specific solutions and maintenance difficulties for large knowledge bases), many researchers from the Artificial Intelligence (AI) community shifted their focus to using physic laws as models to perform diagnosis. It was the time of "reasoning from structure and behavior" (Davis 1982), and the beginning of a new collection of methods.

The AI community has provided different approaches to perform model-based diagnosis (Hamscher *et al.* 1992; Poole 1991), consistency-based diagnosis (CBD for short) being the most popular one. CBD has deep first-order logic roots that allowed the theoretical characterization of its results (Reiter 1987). In the beginning, component-oriented, combinatorial or static and time-invarying systems were the focus of the diagnosis characterization. However, once the theory was established it was immediately extended to dynamic, hybrid, and distributed domains. However, for these kinds of systems there is no formal characterization in first-order logic, but a collection of proposals that follow the main ideas behind the theory first proposed by Reiter (1987):

– contrary to the proposals in traditional FDI theory, the system is made up of a collection of components; the global model is obtained by the compositional modeling of the models of its constituents;

– only information about the system structure (the set of components and their interconnections) and component behavior models can be used to reason about the system state;

– only models of correct (i.e. fault free) behavior are required;

– discrepancies between expected and observed behaviors are used to incriminate components as potential fault candidates. That is, there is no *exoneration*: assuming that a component is correct because there is no evidence of faulty behavior.

There was an initial parallel development of both the theoretical principles and their implementation, the GDE or General Diagnostic Engine (de Kleer and Williams 1987), leading to the first domain independent framework capable of performing multiple-fault diagnosis using only correct behavior models.

6.4.1. *Formalization of consistency-based diagnosis and its first implementation, GDE*

6.4.1.1. *The theoretical framework*

CBD relied upon the *no function in structure* principle: the behavior of each component is represented by a local model that makes no assumption about its working environment: presence or state of another component, operation point, etc. Local models only include relations among the input and output variables at the component terminals (because the model is component-oriented) and, perhaps, the internal parameters of the component. In this way, component behavior does not depend on the environment, and any deviation in the expected behavior can be traced back to the models used to predict the correct behavior, which is explicitly stated as a *correctness assumption* in first-order logic. Additionally, they foster model reuse and the development of model libraries which can be used to generate models for component-oriented systems.

Another novelty in the approach is that the component models are considered constraints among system variables, without imposing any causality among them.

Finally, to allow logical reasoning to solve the diagnosis problem, the model for each component will follow the following pattern:

$$ANDgate(c) \wedge \neg AB(c) \rightarrow Output(c) = Input1(c) \wedge Input2(c)$$

This is the local model for a combinatorial AND-gate, capable of performing the AND operation. If the component is behaving without fault (i.e. it is not *AB*normal) the relation between the inputs and output of the gate holds.

Anytime the diagnosis engine uses the model for a component c, it will record the correctness assumption ($\neg AB(c)$). When a discrepancy is found between a predicted and observed variable, the system can trace back the dependencies and signal out the set of potential fault candidates. For this reason, the theorem prover or inference engine normally used to obtain new deductions, needs to be coupled with a dependency-recording engine.

In the seminal paper *"A theory of diagnosis from first principles"*, Reiter (1987) proposed a general theory in first-order logic that defined the CBD framework. The theory is based on important concepts such as component, observation, system, conflict and diagnosis. All of the definitions are made up of a collection of first-order logic sentences and constants.

DEFINITION 6.1 (System).– *A system is a triplet* $(\mathcal{SD}, COMPS, OBS)$ *where:*

– SD*, the system description, is a finite set of first order sentences.*

– $COMPS$*, the system components, is a finite set of symbols of constants.*

– OBS*, the system observations, is a finite set of first order sentences.*

In Reiter's theory, a *system* specifies a diagnosis problem. SD is the system structure, made up of correct behavior models ($\neg AB\,(x)$) of the set of components, $COMPS$. The diagnosis cannot be carried out without observations, OBS.

We first need to define $D(C_p, C_n)$, which is a conjunctive clause which declares that the set of components in C_p are faulty and the set of components in C_n are non-faulty.

DEFINITION 6.2 ($D(C_p, C_n)$).– *Given two sets of components,* C_p *and* C_n*,* $D(C_p, C_n)$ *is defined to be the conjunction:* $\left[\wedge_{c\in C_p} AB(c)\right] \wedge \left[\wedge_{c\in C_n} \neg AB(c)\right]$

DEFINITION 6.3 (Diagnosis).– *Let* $\Delta \subseteq COMPS$. $D\,(\Delta, COMPS - \Delta)$ *is a diagnosis for* $(SD, COMPS, OBS)$ *if and only if* $SD \cup OBS \cup \{D\,(\Delta, COMPS - \Delta)\}$ *is satisfiable.*

A *diagnosis* is a solution to the diagnosis problem specified by the *system*. If we remove the components in Δ from $COMPS$, the consistency – from a logical point of view – of the *system* is restored.

For practical reasons, we only want the minimal number of fault candidates that suffice to explain all abnormal behaviors. As a result, CBD only looks for the minimal set of components that explains all of them: the set of minimal diagnoses.

DEFINITION 6.4 (Minimal diagnosis).– *Let* $\Delta \subseteq COMPS$. *A diagnosis* $D\,(\Delta, COMPS - \Delta)$ *is a minimal diagnosis for* $(SD, COMPS, OBS)$ *if and only if for no proper subset* Δ' *of* Δ*,* $D\,(\Delta', COMPS - \Delta')$ *is a diagnosis.*

To fully characterize the set of diagnoses for a system, we need the following hypothesis, which only holds if we employ models of correct behavior and no exoneration:

HYPOTHESIS 6.1.– (*Minimal diagnosis hypothesis*) If $D\,(\Delta, COMPS - \Delta)$ is a minimal diagnosis, then for every Δ' superset of Δ, $D\,(\Delta', COMPS - \Delta')$ is also a diagnosis.

Keep in mind that if we extend Reiter's basic theory with models of faulty behavior, or if we use observations to exonerate components, the *minimal diagnosis hypothesis* may no longer hold.

Finding the set of minimal diagnoses, or minimal candidates, we obtain a parsimonious representation of the solution space. To find those minimal candidates, the concept of conflict was introduced by de Kleer and Williams (1987).

DEFINITION 6.5 (AB-literal).– *An AB-literal is one of* $AB\,(c)$ *or* $\neg AB\,(c)$ *for some* $c \in COMPS$.

DEFINITION 6.6 (AB-clause).– *An AB-clause is a disjunction of AB-literals containing no complementary pair of AB-literals.*

DEFINITION 6.7 (Conflict).– *A conflict for* $(SD, COMPS, OBS)$ *is an AB-clause entailed by* $SD \cup OBS$. *A positive conflict is a conflict containing only positive AB-literals.*

Conflicts are a key concept in Reiter's theory too – and in any other model-based diagnosis approaches.

Defining a positive conflict as an AB-clause provides a very intuitive interpretation of a conflict: a conflict identifies a set of components such that at least one of them is faulty – a necessary condition to make the corresponding AB-clause true when every AB-literal is positive. As a result, a conflict can also be interpreted as a set of assumptions of correct behavior that support a symptom leading to an inconsistency. This is the interpretation provided by the GDE and its main computation operation (de Kleer and Williams 1987). GDE is the *defacto* computational reference for CBD, as we will demonstrate later.

For academic purposes, to illustrate the performance on CBD we will use the two-bits full-adder example in Figure 6.1. This is a component-oriented, combinatorial system, whose system model can be made up of a collection of first-order formula. In the next sections we will illustrate the behavior of CBD for dynamic systems, however the theoretical basics remain the same.

The full adder system can be written using a first-order logic formula as:

$$
\begin{aligned}
COMPS &= \{\text{X1}, \text{X2}, \text{A1}, \text{A2}, \text{O1}\}\\
SD &= \{XORgate(\text{c}) \wedge \neg AB(\text{c}) \Rightarrow Output(\text{c}) = Input1(\text{c}) \oplus Input2(\text{c}),\\
&\quad ANDgate(\text{c}) \wedge \neg AB(\text{c}) \Rightarrow Output(\text{c}) = Input1(\text{c}) \wedge Input2(\text{c}),\\
&\quad ORgate(\text{c}) \wedge \neg AB(\text{c}) \Rightarrow Output(\text{c}) = Input1(\text{c}) \vee Input2(\text{c}),\\
&\quad XORgate(\text{X1}), XORgate(\text{X2}), ANDgate(\text{A1}), ANDgate(\text{A2}),\\
&\quad ORgate(\text{O1}),\\
&\quad Input1(\text{A1}) = Input1(\text{X1}), Input2(\text{A1}) = Input2(\text{X1}),\\
&\quad Output(\text{X1}) = Input1(\text{X2}), Output(\text{X1}) = Input1(\text{A2}),\\
&\quad Output(\text{A2}) = Input1(\text{O1}), Output(\text{A1}) = Input2(\text{O1})\},\\
&\quad \text{for c} \in COMPS\\
OBS &= \{Input1(\text{X1}), Input2(\text{X1}), Input2(\text{X2}), Output(\text{X2}), Output(\text{O1})\}
\end{aligned}
$$

The output of gates $X2, O1$ must be the binary values for the sum of inputs for $X1$ and the previous carry $Input_1(X2)$. For instance, given the observations $\{Input1(X1) = 1, Input2(X1) = 1, Input2(X2) = 1\}$ as inputs, the correct result should be $\{Output(X2) = 1, Output(O1) = 1\}$, that is, if we add 1 plus 1 plus 1 in binary, the expected output is 11, which is 3 in decimal basis.

What would be the output of the diagnosis algorithm proposed by Reiter? The empty diagnosis makes the set of formulas satisfiable, so that is the minimal diagnoses set.

What would happen if we observe $\{Output(X2) = 0, Output(O1) = 1\}$ instead?

– Assuming $\neg AB(X1)$ is true and $Input1(X1) = 1$, and $Input2(X1) = 1$, we can infer $Output(X1) = 0$.

– Assuming $\neg AB(X2)$ is true and $Input1(X2) = Output(X1)$, and $Input2(X2) = 1$, we can infer $Output(X2) = 1$.

– This inference is not consistent with the observation for $Output(X2) = 0$.

Hence, either $\neg AB(X1)$ or $\neg AB(X2)$ or both must be wrong. In CBD terminology, $\{X1, X2\}$ is a conflict. Additionally it is a minimal conflict.

DEFINITION 6.8 (Minimal Conflict).– *A conflict π for $(SD, COMPS, OBS)$ is a minimal conflict if no proper subset of π is also a conflict.*

Once we have minimal conflicts we need to provide a mean to compute a minimal diagnosis. Reiter's proposal provides the definition and the algorithm to compute them:

DEFINITION 6.9 (Hitting set).– *Let C be a collection of sets. A hitting set for C is a set $\mathcal{H}$ such that $\mathcal{H} \subseteq \bigcup_{S \in C} S$ and $\forall S \in C$, $\mathcal{H} \cap S \neq \emptyset$. $\mathcal{H}$ is a minimal hitting set for C if no proper subset of $\mathcal{H}$ is a hitting set for C.*

THEOREM 6.1.– (Reiter's characterization of diagnosis). Let $\Delta \subseteq COMPS$, $D(\Delta, COMPS - \Delta)$ is a minimal diagnosis for $(SD, COMPS, OBS)$ if and only if Δ is a minimal hitting set for the collection of minimal conflicts of $(SD, COMPS, OBS)$.

Nevertheless, it should be noted that computing minimal hitting-sets demands great computational effort for large collections of sets. Fortunately, several proposals exist to compute the minimal hitting-sets incrementally, which is very convenient in the diagnosis framework because minimal conflicts do not usually appear at the same time, but appear sequentially.

We now have the tool to compute minimal diagnoses from minimal conflicts: $\{X1\}$ and $\{X2\}$ are minimal diagnoses for the set of minimal conflicts in the previous scenario.

Additionally, in the theorem prover required by Reiter:

– assuming $\neg AB(X2)$ is true and the observation $Output(X2) = 0$ and the observation $Input1(X2) = 0$, we can infer that $Input2(X1) = 0$, using a non-causal reasoning step;

– this inference is not consistent again, and we would obtain the same conflict $\{X1, X2\}$.

There is no other conflict in the system.

If the set of observations had been $\{Input1(X1) = 1, Input2(X1) = 1, Input2(X2) = 1, Output(X2) = 0, Output(O1) = 0\}$, then we would have obtained the very same conflict $\{X1, X2\}$, but additionally:

– assuming $\neg AB(X1)$ is true and $Output(X1) = 0$, then;

– given that $Input1(A2) = Output(X1)$, and assuming $\neg AB(A2)$ is true and $Input2(A2) = 1$, we can infer $Output(A2) = 0$;

– assuming $\neg AB(A1)$ is true and $Input1(A1) = 1$, and $Input2(A1) = 1$, we can infer $Output(A1) = 1$;

– since $Input1(O1) = Output(A2)$, and assuming $\neg AB(O1)$ is true, we can infer that $Output(01) = 1$;

– this inference is not consistent with the observation for $Output(O1) = 0$.

Hence the set of correctness assumptions $\{X1, A2, A1, O1\}$ is another conflict. The set of minimal candidates will be $\{X1\}$, $\{A2\}$, $\{A1\}$, $\{O1\}$ for this conflict alone. But since we have a previous minimal conflict, the set of minimal diagnosis is the minimal hitting set for these two minimal conflicts: $\{X1\}$, $\{X2, A2\}$, $\{X2, A1\}$, and $\{X2, O1\}$. That is, we have a singleton as a possible diagnosis candidate, and three double fault diagnosis candidates.

6.4.1.2. *The computational framework: GDE*

Reiter's theory provides a solid theoretical framework to compute minimal diagnosis candidates from the set of minimal conflicts, but does not provide a specific framework to compute the set of minimal conflicts. Using a simple theorem prover to deduce the set is not feasible for complex models, made up of hundreds of formulas. The required computational framework was produced in parallel to Reiter's formalization: the General Diagnosis Engine, or GDE, proposed by de Kleer and

Williams (1987), which was the first automated diagnosis system capable of diagnosing multiple faults.

GDE provides the diagnosis candidates for those systems that fulfill the requirements of CBD stated in section 6.4. Additionally, the only faults that can be addressed are those related to component malfunctioning, contrary to the idea of faults related to system parameters in FDI. If we want to include faults related to sensors, we need to model the sensor as a component.

The GDE approach to model-based diagnosis uses evidence (i.e. observations) only to reject inconsistent modes, but they are not used to confirm neither a correct nor an abnormal behavior, that is, there is no exoneration.

GDE works incrementally in the following process:

– generation of every possible prediction from the available observations to obtain the global state of the system;

– detection of every symptom, that is, discrepancies between predictions and observations, or between two different predictions for the same variable;

– identification of the minimal conflicts related to each discrepancy;

– incremental generation of minimal diagnosis candidates from the set of minimal conflicts.

Another important issue in behavior estimation is that the local models use all of the information available at the component terminals, and try to propagate (infer new values) using all of the constraints in the model, in whatever direction (i.e. the inference or estimation is made not only in a causal way). Local predictions in each component are propagated through every possible connection, given by the system description.

To achieve an efficient computation of the set of minimal conflicts, the GDE couples its inference engine (which generates estimations) with a *Dependency-Recording Engine*, or DRE, which records every correctness assumption present in the models (the $\neg AB(\cdot)$ predicate). In that way, whenever a discrepancy is found, the DRE has already recorded them, and the generation of the minimal conflicts is automatic. GDE uses an *Assumption-based Truth Maintenance System*, or ATMS, also developed by deKleer and Williams (1987) and de Kleer (1986a, 1986b, 1986c), as its DRE.

The ATMS builds an implicit graph online, adding a new edge every time it performs an inference. Every inference generates a *datum* node, whose parents are other nodes in the graph belonging to different types: *facts*, considered true, for instance, observations; *assumed*, whose truth value is not known but is assumed true in order to do the inference, for instance, the $\neg AB(c)$ assumption produces the $\{c\}$ label for those nodes; and *derived* nodes, which are the new datum obtained from the inference.

The set of nodes used to perform an inference is called the *justification* of the inferred datum. The set of labels, assumed to be true, related to a chain of justifications is called the *environment*. An ATMS *label* is the set of environments where the datum is true. The ATMS guarantees that every environment is minimal.

Every time the inference engine obtains a new datum, the ATMS is responsible for computing its environment. When there is a discrepancy for a datum (there are two different values estimated or observed), the ATMS finds the set of labels used for the estimation. This is the source of a minimal conflict, which is represented as a falsity node – whose environment is called NOGOOD – with an empty assumption.

It should be rather easy for the attentive reader to translate the steps performed in the previous section to compute minimal diagnosis for the full-adder example when the set of observations is $OBS = \{Input1(X1) = 1,\ Input2(X1) = 1,$ $Input2(X2) = 1,\ Output(X2) = 0,\ Output(O1) = 0\}$:

1) Facts, $Input1(X1) = 1$ and $Input2(X1) = 1$ with labels $\{\}$, and assumed node $\{X1\}$, derives $Output(X1) = 0$ with label $\{\{X1\}\}$.

2) Fact, $Input2(X2) = 1$ with label $\{\}$, together with derived node $Output(X1) = 0$ with label $\{\{X1\}\}$, and assumed node $\{X2\}$ with label $\{\{X2\}\}$, derives $Output(X2) = 1$ with label $\{\{X1, X2\}\}$.

3) Fact $Output(X2) = 0$ with label $\{\}$, together with the previous datum, generates a NOGOOOD node with label $\{\{X1, X2\}\}$. Hence, $\{\{X1, X2\}\}$ is a minimal conflict.

Since this is the first minimal conflict, the minimal hitting set provides the set of minimal candidates: $\{\{X1\}, \{X2\}\}$.

4) Fact, $Input2(X2) = 1$ with label $\{\}$, together with derived node $Output(X1) = 0$ with label $\{\{X1\}\}$, and assumed node $\{A2\}$ with label $\{\{A2\}\}$, derives $Output(A2) = 0$ with label $\{\{X1, A2\}\}$.

5) We assume in this example that sentences $Input1(\text{A1}) = Input1(\text{X1})$, and $Input2(\text{A1}) = Input2(\text{X1})$ introduce no new node or label.

Facts, $Input1(A1) = 1$ and $Input2(A1) = 1$ with labels $\{\}$, and assumed node $\{A1\}$, derives $Output(A1) = 1$ with label $\{\{A1\}\}$.

6) Derived nodes $Output(A1) = 1$ with label $\{\{A1\}\}$, and $Output(A2) = 0$ with label $\{\{X1, A2\}\}$, and assumed node $\{O1\}$ with label $\{\{O1\}\}$, derives $Output(O1) = 1$ with label $\{X1, A2, A1, O1\}$.

7) Fact $Output(O1) = 0$ with label $\{\}$, together with the previous datum, generates a NOGOOOD node with label $\{\{X1, A2, A1, O1\}\}$. Hence, $\{\{X1, A2, A1, O1\}\}$ is the second minimal conflict.

Using an incremental approach, the minimal hitting set for this minimal conflict and the two previous minimal diagnosis candidates give rise to the following set of minimal candidates: $\{\{X1\}$, as a singleton because it is a member of the two minimal conflicts, and the set of double fault candidates: $\{\{X2, A2\}, \{X2, A1\}, \{X2, O1\}\}$.

As expected, these are the same results that were obtained by applying Reiter's framework.

The ability of the inference engine to propagate both in the causal direction, from inputs to output, or vice versa provides other *nogood* nodes, for instance starting for output observations for $X2$ and $O1$ and going backwards, but all of them are equivalent to the two previous ones and they are not recorded. This fact increases efficiency while reducing space memory.

There is another step in the iterative process to generate minimal candidates in GDE, which is the selection of new observations that may help to reduce the set of consistent candidates. This process is based on information theory, looking for the next observation that maximizes information gain. However, due to a lack of space and because in real systems usually there is a small set of observations and there is no chance to ask for new measurements, we recommend the interested reader consult the original paper (de Kleer and Williams 1987), or the summary provided in Chapter 5 of Escobet *et al.* (2019), for further details.

6.4.1.3. *Computational alternatives to GDE*

Consistency-based diagnosis theoretical characterization using first-order logic (Reiter 1987) and GDE, its computational *defacto* paradigm (de Kleer and Williams 1987), encountered several problems when faced with real-life complex dynamic systems:

1) The ATMS used for dependency-recording in GDE needs to store every new inference, which gave rise to memory problems and produced computational overload for real-valued domains. This problem is even bigger with dynamic models, either quantitative or qualitative, used for simulation because of the correctness and time dependencies for each state variable with its values at previous time steps.

2) There is no extension of Reiter's theory for dynamic systems.

The second issue is closely related to the first. Several systems that tackle the first issue for complex dynamic systems have become the *defacto* references for CBD of dynamic systems.

Several proposals were immediately available to cope with both issues, primarily in the following two ways:

– The first approach was to extend the GDE.

These works provided different extensions, such as work in multiple contexts (Guckenbiehl and Schäfer-Richter 1990), to include time steps as part of the prediction (Dague 1994; Loiez and Taillibert 1997), or to avoid simulation of qualitative models (Dressler 1996). This last method is also known as state-based diagnosis (Struss 1997). All of them work for qualitative or discrete domains, but they are not able to work with traditional engineering analytical models.

– The second approach was to avoid on-line dependency-recording, as performed by the ATMS.

This approach is also known to be based on topolocial methods, because they use some kind of graphical representation ((temporal) causal graphs, bond-graphs, bipartite graphs or hypergraphs) related to the numerical models (each equation or relation provides an edge between their variables (nodes)). Conflicts are searched for in these graphs instead of using an ATMS for each inference. The graphs are explored either online or offline, either backward or forward, from inputs to outputs or vice versa. The paths between observed values used as inputs and the observed outputs leading to a discrepancy define a subgraph which is considered a conflict related to that discrepancy. At least one of those edges (related to equations) with their parameters (related to faults) must be wrong for the discrepancy to be present.

Some of these topological methods, which perform the search online, have become the *defacto* references for CBD of dynamic systems, such as Ca-En (Travé-Massuyès 2006) or TRASCEND (Mosterman and Biswas 1999). Other methods, that perform the search offline (Pulido and Alonso-González 2004; Krysander *et al.* 2008; Armengol *et al.* 2009; Bregon *et al.* 2014b), are closely related to the structural analysis methods described in Chapter 2.

Since all of these topological methods have been designed for dynamic systems diagnosis, we will illustrate the behavior of one of them for the three-tanks case study in the next section.

6.5. CBD for dynamic systems

Fault diagnosis of dynamic systems has, since the beginning of the DX approach, been a major research topic within the community. In 1984, even before Reiter's theory and the GDE proposal, the work by Hamscher and Davis (1984), proposed to extend constraint suspension techniques to deal with dynamic models. This work also highlighted the need to observe the system state variables in order to be able to provide deterministic results.

Due to a lack of space it is impossible to summarize the set of different approaches existing in the AI community. The main problem to reason about system dynamics within this community is the absence of a formal and unique theory about the modeling of temporal information, existing definitions for concepts such as time, temporal behavior, temporal dynamics, or even diagnosis itself. (Brusoni *et al.* (1998) and references therein) provided an excellent summary of all these problems and suggested a set of dimensions to categorize existing approaches as of 1998. Their main conclusion is that instead of having a general theory to extend Reiter's proposal to dynamic systems, as it happens in the FDI community, the DX community has evolved from a collection of separate works as different references for the community.

Those dimensions are:

1) **Temporal phenomena**: which refers to the temporal aspects in the system, such as *time-varying contexts* (observations at different times where the system might have changed), *temporal behavior* (state changes over time in a discrete or continuous way), and *time-varying behavior* (component modes can change over time, from normal to faulty and vice versa; for instance, allowing intermittent faults or not).

2) **Ontologies for time**: *metric time* (i.e. quantitative models as in the FDI community), *qualitative notion of temporal ordering* as in Allen's interval algebra (Allen 1983), or other ad-hoc time modeling approaches.

3) **Diagnosis**: there are different ways to understand temporal diagnosis regarding its "explanation capability of temporal observations": *consistency-based diagnosis* (which aims for diagnosis that are consistent, i.e. not refuted by observations), or *abductive diagnosis* (which aims for a complete model capable of explaining all available observations).

For the sake of simplicity and looking for system models and diagnosis proposals easily comparable to FDI concepts, such as dynamic models made up of Ordinary Differential Equations, we will focus on:

1) time-varying behavior systems;

2) with continuous state and parameter changes over time;

3) with no intermittent faults allowed;

4) using a quantitative metric time;

5) using consistency-based diagnosis.

Within this framework, the main issue in modeling dynamics comes from the representation of the system state variables, and the influence in the current state from previous states (assuming that the diagnosis of combinatorial circuits, i.e. with no state, can be seen as a sequence of static diagnosis processes). As pointed out by Chantler *et al.* (1996), in quantitative terms, the relation in discrete time should be able to be represented as:

$$x(t) = x(t-1) + \Delta t \frac{dx(t-1)}{dt}$$

This kind of relation is usually named a *differential* constraint, while algebraic or functional relations in the model are usually named *instantaneous* constraints.

If the term $\frac{dx(t-1)}{dt}$ can be computed, then integration can be performed and $x(t)$ obtained by simulation. If the state variables $x(t)$ can be observed, $\frac{dx(t)}{dt}$ can be computed and later estimate the next state as $x(t+1) = x(t) + \Delta t \frac{dx(t)}{dt}$.

6.5.1. *Different approaches for CBD of dynamic systems*

As previously mentioned, coupling complex logic-based dynamic system models with Dependency-Recording Engines, such as ATMS, was a major problem for consistency-based diagnosis. There were two major alternatives: first, to extend GDE for dynamic systems; second, to use other alternatives to online DREs.

The GDE framework was extended to work with inferences, including temporal information that has given rise to different diagnosis systems: Guckenbiehl and Schäfer-Richter (1990); Dague *et al.* (1990); Loiez and Taillibert (1997); Dressler (1996). But they kept several drawbacks:

1) the need to remove dependencies out of a recent temporal window to save memory;

2) the need to change the probabilistic tests to select new probe points for diagnosis candidates refinement;

3) there was no direct extension to Reiter's theory for dynamic systems to logically support all of the results.

Furthermore, the relation between current and past values for state variables was still there. Some researchers decided to separate inferences at the same time step (instantaneous or intra-state) to those made between consecutive time steps (differential or inter-state) (Struss 1997). The proposal by Struss *et al.* was to only look at intra-state estimations to find conflicts, and it defined an approach that is known as state-based diagnosis (categorized in Brusoni *et al.* 1998), which only works for qualitative models under given assumptions. But mostly, DX approaches rely upon some kind of coupling between a model used for behavior simulation and a parallel graphical model used to find conflicts. These are named **topological** approaches to CBD and they will be the focus of our discussion.

The main idea behind topological methods is that the system structure and the set of available measurements will be fixed most of the time. Hence, the propagation of information in the system will be limited by the topology of the system (how components are connected) and measurement availability.

Topological methods can be later divided into online or offline dependency-recording. The first category explores the graphical structure looking for conflicts once a discrepancy is found using the quantitative models. The second category performs an offline exploration of the graphical structure to find the whole set of possible minimal conflicts beforehand, in a similar way to the structural methods previously presented in this book.

Two of the most well-known topological methods that perform online searches of conflicts are: Ca∼En (Bousson *et al.* 1994; Milne and Travé-Massuyès 1997) and TRANSCEND (Mosterman and Biswas 1999; Manders *et al.* 2000). Consistency-based diagnosis using Possible Conflicts (Pulido and Alonso-González 2004; Pulido *et al.* 2001; Bregon *et al.* 2012), is one of the well known techniques using offline dependency-recording. A complete summary of these approaches is provided in Chapter 6 of Escobet *et al.* (2019).

– Ca∼En proposal for causal reasoning has been used for diagnosis combining two different types of models: a local model (a causal graph where system variables were connected through links known as influences) and a global model (consisting of quantitative or numerical interval relations among system variables). Influences can be used to model static or dynamic relations among variables. The global model is made up of nonlinear intervalar models that can be checked for residual generation. Once a residual is found, the local model related to those variables in the global model is checked for consistency. When a inconsistency is found in the model, those influences in the path to the inconsistency are considered a conflict.

– The TRANSCEND proposal starts from a bond-graph model (Broenink 1999) of the system. Bond-graphs, BG for short, are general, domain independent models that represent relations between effort (such as voltage or pressure) and flow (such as electric current or fluid flows) variables in different domains. The main advantage of BGs is that different types of causal graphs and equations can be derived automatically, because each edge in the BG model can be translated to one analytical equation or to a causal relation among effort and flow variables.

TRANSCEND uses a numerical, real-valued state-space model which is used to design an Extended Kalman Filter (EKF) which tracks the system observed variables, and to generate a set of residuals. The numerical values of the residuals and their derivatives are translated into qualitative values such as $\{-, 0, +\}$. For instance the pair $\{+, 0\}$ represent a qualitative value which is above its expected value and it is in stationary state (its derivative is 0).

Additionally, derived from the BG model, TRANSCEND has a temporal causal graph, which relates system variables and parameters. It is used to reason about the consistency of the qualitative values obtained in the previous stage. The values are propagated backward in the graph looking for inconsistencies. When an inconsistency is found, all of the parameters in the edges traversed in that search are selected as hypothetical faults. For each fault the graph is traversed forward and their hypothetical signatures (qualitative effects on observed variables) are generated. The consistency of these signatures is checked against real observations, rejecting those faults whose signature is not consistent with observations. Finally, for those faults consistent with observations, an identification process by means of numerical optimization is performed.

– CBD with Possible Conflicts (PCs), is a system decomposition method from the DX community, used to find the set of minimal conflicts in a system model offline. Once the set of PCs is found, they can be implemented using different types of numerical models: DAEs, EKF, Neural Networks, or Dynamic Bayesian Networks.

Given a set of ODEs or DAEs as a system description, in order to find all of the subsystems capable of becoming conflicts, as in the GDE framework, it is necessary to perform a model abstraction. The abstraction generates an hypergraph (Murota 2012), whose nodes are system variables, and whose hyperarcs represent the relations between the variables in the model. Regarding the nodes, that is, system variables, it is important to distinguish between observed and non-observed variables.

For instance, equation $e1$ generates a hyperarc: $(e1 : \{\dot{h}_{T_1}, q_i, q_{12}, q_{f_1}\}$. In this hyperarch only q_i is observed. The remainder variables (nodes) are not observed.

The next step is to perform an exhaustive depth-first search in the graph from each variable in order to find the subsystems with minimal analytical redundancy. Each one of these partial hypergraphs is called a *Minimal Evaluable Chain* (MEC), and they are equivalent to minimal Analytical Redundant Relations. The search stops when the number of equations is exactly one more that the number of non-observed variables.

For instance, by searching for a minimal redundant subsystem starting from h_{T_3}, we find the following MEC: $\{e3, e8, e9, e12\}$. To model the system dynamics we add one additional constraint that relates each state-variable and its derivative. In the three-tank example, we have three state-variables $\{h_{T_1}, h_{T_2}, h_{T_3}\}$, and their derivatives $\{\dot{h}_{T_1}, \dot{h}_{T_2}, \dot{h}_{T_3}\}$. These equations are numbered as $\{e10, e11, e12\}$, respectively. We have also assumed that variables $\{q_i, h_{T_1}, q_{23}, h_{T_3}\}$ are the only observed variables. Finally, $\{q_{12}, q_{30}, q_{f_1}, q_{f_2}, q_{f_3}, \dot{h}_{T_1}, h_{T_2}, \dot{h}_{T_2}, \dot{h}_{T_3}\}$ are the set of unknown variables.

Minimal Evaluable Chains represent sets of equations with structural redundancy, but it is not possible to know if the set of equations represented by the MEC is solvable or not. To do so, a directed and-or graph is built for each MEC. Each hyperarc in the MEC provides a set of one or more edges in the and-or graph: one for each available causal assignment.

For instance, edge $e3$, can only be solved as shown in equation $e3$: the derivative for h_{T_3} is computed from q_{23}, q_{30} and q_{f_3}. The same can be applied for all of the equations $e1$ to $e9$. But for inter-state constraints such as $\{e10, e11, e12\}$, we can have two causal assignments. $e12_1$ would represent the computation of h_{T_3}, given its derivative $\dot{h}_{T_3}$. $e12_2$ would represent the computation of the derivative $\dot{h}_{T_3}$, given h_{T_3}. And so on.

As a result, each MEC can have zero, one or more ways of solving the underlying set of equations with respect to the set of available causal assignments. Each way of solving the equations is called a consistent global causal assignment for the equations in a MEC and it is termed a *Minimal Evaluable Model* or MEM. In this case for MEC $\{e3, e8, e9, e12\}$ we have at least one MEM: $\{e3_1, e8_1, e9_1, e12_1\}$.

All of these computations are made offline and no model is evaluated. Once the offline stage is over, each MEC can be implemented as a simulation model (Pulido *et al.* 2001), as a state-observer – for instance an EKF (Bregon *et al.* 2014a) – or a Dynamic Bayesian Network (Alonso-Gonzalez *et al.* 2010). These models can be used to track system behavior and activate residuals.

PCs have been demonstrated to be equivalent to the set of minimal conflicts (Pulido and Alonso-González 2004), and also to the set of minimal ARRs and MSOs (Armengol *et al.* 2009; Krysander *et al.* 2008).

Whenever a residual that is related to a Possible Conflict is detected, the set of equations in its MEC are used as a minimal conflict, and the usual CBD cycle (behavior prediction, conflict detection, candidate generation) can be followed.

In the next section, we will illustrate this last approach for the three-tank case study.

6.5.2. *PCs for the three-tank system case study*

As mentioned in the previous section, in order to compute the Possible Conflicts we first need to include additional constraints to our model. In particular, we need to include the following set of equations to make the relation between the state variables and their derivatives explicit:

$$e10 : h_{T_1} = \int \dot{h}_{T_1} dt,$$

$$e11 : h_{T_2} = \int \dot{h}_{T_2} dt,$$

$$e12 : h_{T_3} = \int \dot{h}_{T_3} dt.$$

In this example, since we are not including sensor faults, it makes no sense to make an explicit difference between an observed variable and its corresponding internal variable. To make clear which are the observed variables, we have added an asterisk to each one of them: $\{q_i^*, q_{23}^*, h_{T_1}^*, h_{T_3}^*\}$.

To obtain the hypergraph related to the system model we just need to add one hyperarc for each DAE in the model. The set Σ_{3Ts} provides the set of hyperarcs $\mathcal{R}_{3Ts}$:

$$\begin{gathered}
(e1 : \{\dot{h}_{T_1}, q_i, q_{12}, q_{f_1}\}) \\
(e2 : \{\dot{h}_{T_2}, q_{12}, q_{23}, q_{f_2}\}) \\
(e3 : \{\dot{h}_{T_3}, q_{23}, q_{30}, q_{f_3}\}) \\
(e4 : \{q_{12}, h_{T_1}, h_{T_2}\}) \\
(e5 : \{q_{f_1}, h_{T_1}\}) \\
(e6 : \{q_{23}, h_{T_2}, h_{T_3}\}) \\
(e7 : \{q_{f_2}, h_{T_2}\}) \\
(e8 : \{q_{30}, h_{T_3}\}) \\
(e9 : \{q_{f_3}, h_{T_3}\}) \\
(e10 : \{h_{T_1}, \dot{h}_{T_1}\}) \\
(e11 : \{h_{T_2}, \dot{h}_{T_2}\}) \\
(e12 : \{h_{T_3}, \dot{h}_{T_3}\})
\end{gathered}$$

The resulting hypergraph will be $H_{3Ts} = \{\mathcal{V}_{3Ts}, \mathcal{R}_{3Ts}\}$, being $\mathcal{V}_{3Ts} = \{q_i^*, h_{T_1}^*, h_{T_2}, h_{T_3}^*, \dot{h}_{T_1}, \dot{h}_{T_2}, \dot{h}_{T_3}, q_{12}, q_{23}^*, q_{30}, q_{f_1}, q_{f_2}, q_{f_3}\}$, the set of variables and $\mathcal{R}_{3Ts} = \{e1, e2, e3, e4, e5, e6, e7, e8, e9, e10, e11, e12\}$, the set of relations/equations.

In this work we have used the algorithms described in Pulido and Alonso-González (2004) to derive the set of MECs, $H_{MEC_m} = (\mathcal{V}_{MEC_m}, \mathcal{R}_{MEC_m})$, for $m = 1$ to 5, which can be seen in Table 6.1. In this table, $\mathcal{R}_{MEC_m} \subset \mathcal{R}_{3Ts}$ represents the set of relations/equations in MEC_m, and $\mathcal{V}_{MEC_m} \subset \mathcal{V}_{3Ts}$ represents the set of variables in MEC_m (with $\mathcal{Y}_m$, $\mathcal{X}_m$ and $\mathcal{S}_m$ being the observed output, the nonobserved and the observed input variables, respectively).

In the three-tank system, we have provided exactly one potential causal assignment for each constraint, and it is the one shown in equations $e1$ to $e9$. Regarding differential constraints, we have allowed only integral causality, as shown in equations $e10$ to $e12$.

MEC_m	$\mathcal{R}_{MEC_m}$	$\mathcal{V}_{MEC_m} = \mathcal{Y}_m \cup \mathcal{X}_m \cup \mathcal{S}_m$
MEC_1	$\{e3, e8, e9, e12\}$	$\{h^*_{T_3}\} \cup \{h_{T_3}, q_{30}, q_{f_3}\} \cup \{q^*_{23}\}$
MEC_2	$\{e2, e4, e6, e7, e11\}$	$\{q^*_{23}\} \cup \{h_{T_2}, \dot{h}_{T_2}, q_{12}, q_{f_2}\} \cup \{h^*_{T_3}, h^*_{T_1}\}$
MEC_3	$\{e1, e2, e4, e5, e6, e7, e10, e11\}$	$\{h^*_{T_1}\} \cup \{\dot{h}_{T_1}, q_{12}, h_{T_2}, \dot{h}_{T_2}, q_{f_1}, q_{f_2}\} \cup \{q^*_i, q^*_{23}\}$
MEC_4	$\{e1, e4, e5, e6, e10\}$	$\{\} \cup \{h_{T_2}, \dot{h}_{T_1}, q_{12}, q_{23}, q_{f_1}\} \cup \{q^*_i, h^*_{T_1}, h^*_{T_3}\}$
MEC_5	$\{e1, e2, e5, e6, e7, e10, e11\}$	$\{\} \cup \{\dot{h}_{T_1}, \dot{h}_{T_2}, q_{12}, h_{T_2}, q_{f_1}, q_{f_2}\} \cup \{h^*_{T_1}, q^*_{23}, h^*_{T_3}\}$

Table 6.1. *Set of Minimal Evaluable Chains for the three-tank system. Only three of them have a Minimal Evaluable Model and identify a Possible Conflict*

Using these causality assignments we have obtained exactly one MEM for MEC_1, MEC_2, and MEC_3; MEC_4, and MEC_5 have no global valid causal assignment, hence they are not Possible Conflicts. On the other hand, each MEC with one MEM defines a Possible Conflict: $\{PC_1, PC_2, PC_3\}$, capable of estimating variables $\{h^*_{T_3}, q^*_{23}, h^*_{T_1}\}$, respectively.

To summarize, each MEM for a PC represents how to build an executable model to monitor the behavior of the subsystem defined by the PC, just traversing the MEM structure from the leaves to the root (which is $\hat{y}_m$). Such an executable model can be implemented as a simulation model or as a state-observer (see Pulido *et al.* 2010; Bregon *et al.* 2014a). However, building such a model for complex nonlinear systems is not a trivial task.

The software required to perform a simple dynamic system diagnosis, given a system model specified as a set of DAEs and the set of PCs, named *DBCwithPCs*, can be found at: https://www.infor.uva.es/ belar/SoftwareCPCs/

6.6. Conclusion

AI has embarked upon system diagnosis from its very beginning. KBS, CBR, Machine Learning, and model-based systems are the four main ways the field has provided different solutions to such complex problems. In this chapter we have reviewed two of them: knowledge-based and model-based systems.

KBS are no longer a research field for diagnosis, but now provide many different practical systems in several domains. Their main drawbacks are that they provide device-dependent solutions, have maintenance (and coherence) problems for large knowledge-bases, and have difficulties dealing with multiple or unexpected faults.

Model-based systems started with logically-sound solutions for static, time-unvarying systems. Their main advantages were: domain independent solutions and automatic multiple fault diagnosis with only correct behavior models. The main trend among AI solutions is consistency-based diagnosis, which does not initially require fault models and does not confirm behavior modes – instead, it uses inconsistent observations to reject faulty behaviors. The component-oriented, no-function-in-structure principles fostered the reuse and development of model libraries.

Diagnoses from first principles (either the Reiter's logical formulation and the GDE implementation) were able to provide a parsimonious representation of the set of minimal diagnoses candidate space from the set of minimal conflicts. The need for dependency-recording to keep track of conflicts and their problems when introducing time label annotations provided another topic of research in the field: online vs offline dependency-recording. This issue was even more critical when dealing with temporal information.

Soon after the AI foundations had settled, researchers were also working on extensions to deal with dynamic systems, and so on. There is no general extension to Reiter's diagnosis characterization for dynamic systems, but through the years several proposals have become *de facto* alternatives: topological methods such as Ca-En or TRANSCEND, or offline dependency-compilation such as Possible Conflicts. All of them use a graphical representation of the system model which is explored in order to find conflicts (components whose assumption of correct behavior is not consistent with observations). Additionally, several extensions to the original GDE ideas have been proposed over the last 25 years.

The need to diagnose complex real-life systems (in terms of model complexity or number of components) has prompted the use of AI techniques to provide solutions for discrete event systems, hybrid systems, or autonomous systems. Traditional approaches that rely upon qualitative variables have evolved to include numerical and probabilistic models. Increasingly, most diagnosis systems make use of techniques from the fields of both AI and FDI.

However, there are still some differences between the FDI and the AI approaches to model-based diagnosis:

– AI methods usually pay more attention to the isolation and identification stages of fault diagnosis, mainly using qualitative or symbolic information, instead of paying attention to robust fault detection and residual analysis.

– AI methods usually work at the component level, using compositional modeling, instead of analyzing system models as a whole.

– Consistency-based diagnosis does not allow exoneration (to confirm a correct or fault mode based on current observations), instead it relies upon rejecting those modes which are not consistent with observations.

The need for hybrid solutions, however, taking the best from the AI and the FDI worlds, is real and is paving the way for future solutions.

6.7. References

Allen, J.F. (1983). Maintaining knowledge about temporal intervals. *Communications of the ACM*, 26(11), 832–843.

Alonso-Gonzalez, C.J., Moya, N., Biswas, G. (2010). Factoring dynamic Bayesian networks using possible conflicts. In *Proc. of the 21th International Workshop on Principles of Diagnosis*, 7–14.

Armengol, J., Bregon, A., Escobet, T., Gelso, E., Krysander, M., Nyberg, M., Olive, X., Pulido, B., Travé-Massuyès, L. (2009). Minimal structurally overdetermined sets for residual generation: A comparison of alternative approaches. In *Proceedings of the IFAC-SafeProcess'09*, Barcelona, Spain.

Balakrishnan, K. and Honavar, V. (1998). Intelligent diagnosis systems. *Journal of Intelligent Systems*, 8(3/4), 239–290.

Bousson, K., Travé-Massuyès, L., Zimmer, L. (1994). Causal model-based diagnosis of dynamic systems. Report, LAAS, No. 94231.

Bregon, A., Biswas, G., Pulido, B. (2012). A decomposition method for nonlinear parameter estimation in TRANSCEND. *IEEE Transactions on Systems, Man, and Cybernetics-Part A: Systems and Humans*, 42(3), 751–763.

Bregon, A., Alonso-González, C.J., Pulido, B. (2014a). Integration of simulation and state observers for online fault detection of nonlinear continuous systems. *IEEE Transactions on Systems, Man, and Cybernetics: Systems*, 44(12), 1553–1568.

Bregon, A., Biswas, G., Pulido, B., Alonso-González, C., Khorasgani, H. (2014b). A common framework for compilation techniques applied to diagnosis of linear dynamic systems. *IEEE Transactions on Systems, Man, and Cybernetics: Systems*, 44(7), 863–876.

Broenink, J. (1999). Introduction to physical systems modeling with bond Graphs. *SiE Whitebook on Simulation Methodologies*, 31(2).

Brown, J., Burton, R., DeKleer, J. (1982). Pedagogical and knowledge engineering techniques in SOPHIE I, II and III. In *Intelligent Tutoring Systems*, Sleeman, D., Brown, J.S. (eds), Academic Press, London.

Brusoni, V., Console, L., Terenziani, P., Dupré, D.T. (1998). A spectrum of definitions for temporal model-based diagnosis. *Artificial Intelligence*, 102(1), 39–79.

Buchanan, B.G. and Shortliffe, E.H. (eds) (1984). *Rule-Based Expert Systems, The MYCIN Experiments of the Stanford Heuristic Programming Project*. Addison Wesley, Publishing Company, Boston, MA.

Chantler, M., Daus, S., Vikatos, T., Coghill, G. (1996). The use of quantitative dynamic models and dependency recording engines. In *Proceedings of the 7th International Workshop on Principles of Diagnosis (DX-96)*, Val Morin, Quebec, Canada, 59–68.

Clancey, W.J. (1985). Heuristic classification. *Artificial Intelligence Journal*, 27(3), 289–350.

Console, L. (2000). Model-based diagnosis history and state of the art. *MONET Summer School*, Bertinoro, Italy.

Dague, P. (1994). Model-based diagnosis of analog electronic circuits. *Annals of Mathematics and Artificial Intelligence*, 11(1), 439–492 [Online]. Available at: https://doi.org/10.1007/BF01530755.

Dague, P., Deves, P., Luciani, P., Taillibert, P. (1990). Analog systems diagnosis. In *Proceedings of European Conference on Artificial Intelligence (ECAI-90)*, 173–178.

Daniele, L.M., Punter, M., Brewster, C., García-Castro, R., Poveda, M., Fernández, A. (2018). *A SAREF Extension for Semantic Interoperability in the Industry and Manufacturing Domain*. Wiley Online Library, Hoboken, NJ.

Davis, R. (1982). Expert systems: Where are we? And where do we go from here? *Artificial Intelligence*, 3, 3–22.

Ding, S.X. (ed.) (2014). Case study and application examples. In *Data-driven Design of Fault Diagnosis and Fault-tolerant Control Systems*. Springer-Verlag, London

Dressler, O. (1996). On-line diagnosis and monitoring of dynamic systems based on qualitative models and dependency-recording diagnosis engines. In *Proceedings of the 12th European Conference on Artificial Intelligence (ECAI-96)*, 461–465.

Escobet, T., Bregon, A., Pulido, B., Puig, V. (2019). *Fault Diagnosis of Dynamic Systems*. Springer, Cham.

Fernández-Izquierdo, A., García-Castro, R., Daniele, L., de Roode, M. (n.d.). SAREF extension for industry and manufacturing domain [Online]. Available at: https://mariapoveda.github.io/saref-ext/OnToology/SAREF4INMA/ontology/saref4inma.ttl/documentation/index-en.html.

Gertler, J. (1998). *Fault Detection and Diagnosis in Engineering Systems*. CRC Press, Boca Raton, FL.

Guckenbiehl, T. and Schäfer-Richter, G. (1990). SIDIA: Extending prediction based diagnosis to dynamic models. In *Expert Systems in Engineering Principles and Applications*, Gottlob, G. and Nejdl, W. (eds). Springer-Verlag, Berlin, Heidelberg.

Hamscher, W. and Davis, R. (1984). Diagnosing circuits with state: An inherently underconstrained problem. In *Procs. of AAAI*, 142–147.

Hamscher, W., Console, L., de Kleer, J. (1992). *Readings in Model-Based Diagnosis*. Morgan Kaufmann Publishers Inc, San Francisco, CA.

Jackson, P. (1998). *Introduction to Expert Systems*. Addison Wesley Longman Publishing Co., Inc, Upper Saddle River, NJ.

Kahn, G. and McDermott, J. (1986). The mud system. *IEEE Expert*, 1(1), 23–32.

de Kleer, J. (1986a). An assumption-based TMS. *Artificial Intelligence*, 28, 127–162.

de Kleer, J. (1986b). Extending the ATMS. *Artificial Intelligence*, 28, 163–196.

de Kleer, J. (1986c). Problem solving with the ATMS. *Artificial Intelligence*, 28, 197–224.

de Kleer, J. and Williams, B.C. (1987). Diagnosing multiple faults. *Artificial Intelligence*, 32(1), 97–130.

Kolodner, J. (2014). *Case-based Reasoning*. Morgan Kaufmann, Publishers Inc, San Francisco, CA.

Krysander, M., Åslund, J., Nyberg, M. (2008). An efficient algorithm for finding minimal overconstrained subsystems for model-based diagnosis. *IEEE Transactions on Systems, Man, and Cybernetics-Part A: Systems and Humans*, 38(1), 197–206.

Loiez, E. and Taillibert, P. (1997). Polynomial temporal band sequences for analog diagnosis. In *Proceedings of the 15th International Joint Conference on Artificial Intelligence (IJCAI-97)*, Nagoya, Japan, 474–479.

Manders, E.-J., Narasimhan, S., Biswas, G., Mosterman, P.J. (2000). A combined qualitative/quantitative approach for fault isolation in continuous dynamic systems. *Proceedings of the IFAC-SafeProcess*, Budapest, Hungary, 512–517.

Matsatsinis, N., Doumpos, M., Zopounidis, C. (1997). Knowledge acquisition and representation for expert systems in the field of financial analysis. *Expert Systems with Applications*, 12(2), 247–262.

Milne, R. and Travé-Massuyès, L. (1997). Model based aspects of the TIGER gas turbine condition monitoring system. *IFAC Proceedings Volumes*, 30(18), 405–410.

Mitchell, T.M. (2006). *The Discipline of Machine Learning*, vol. 9. Carnegie Mellon University, School of Computer Science, Machine Learning, Pittsburgh, PA.

Mosterman, P. and Biswas, G. (1999). Diagnosis of continuous valued systems in transient operating regions. *IEEE Transactions on Systems, Man, and Cybernetics*, 29(6), 554–565.

Murota, K. (2012). *Systems Analysis by Graphs and Matroids: Structural Solvability and Controllability*, vol. 3. Springer Science & Business Media, Berlin, Heidelberg.

Pauwels, P. and Terkaj, W. (n.d.). ifcOWL ontology (IFC4x1) [Online]. Available at: https://standards.buildingsmart.org/IFC/DEV/IFC4_1/OWL.

Peng, Y. and Reggia, J.A. (1990). *Abductive Inference Models for Diagnostic Problem-Solving*. Springer-Verlag, New York.

Poole, D. (1991). Representing diagnostic knowledge for probabilistic horn abduction. In *IJCAI*, 1129–1137.

Pulido, B. and Alonso-González, C. (2004). Possible conflicts: A compilation technique for consistency-based diagnosis. *IEEE Trans. on Systems, Man, and Cybernetics. Part B: Cybernetics*, 34(5), 2192–2206.

Pulido, B., Alonso, C., Acebes, F. (2001). Consistency-based diagnosis of dynamic systems using quantitative models and off-line dependency-recording. In *12th International Workshop on Principles of Diagnosis (DX-01)*, Sansicario, Italy, 175–182.

Pulido, B., Bregon, A., Alonso-González, C. (2010). Analyzing the influence of differential constraints in possible conflicts and ARR computation. In *Current Topics in Artificial Intelligence, CAEPIA 2009 Selected Papers*, Meseguer, P., Mandow, L., Gasca, R.M. (eds). Springer-Verlag, Berlin.

Reggia, J.A., Nau, D., Wang, P. (1983). Diagnostic expert systems based on a set covering model. *International Journal of Man-Machine Studies*, 19(3).

Reiter, R. (1987). A theory of diagnosis from first principles. *Artificial Intelligence*, 32(1), 57–95.

Shortliffe, E.H. (1976). *Computer-based Medical Consultations: MYCIN*. Elsevier, Amsterdam.

Stefik, M. (2014). *Introduction to Knowledge Systems*. Elsevier, Amsterdam.

Struss, P. (1997). Fundamentals of model-based diagnosis of dynamic systems. In *Proceedings of the 15th International Joint Conference on Artificial Intelligence (IJCAI-97)*, Nagoya, Japan, 480–485.

Travé-Massuyès, L. (2006). Model based thoughts Ca-En and TIGER then and now. In *ROB MILNE: A Tribute to a Pioneering AI Scientist, Entrepreneur and Mountaineer*, Bundy, A., Wilson, S. (eds). IOS Press, Amsterdam.

Venkatasubramanian, V., Rengaswamy, R., Yin, K., Kavuri, S.N. (2003). A review of process fault detection and diagnosis. Part I: Quantitative model-based methods. *Computers & Chemical Engineering*, 27(3), 293–311 [Online]. Available at: http://www.sciencedirect.com/science/article/pii/S0098135402001606.

Watson, I. (1998). *Applying Case-Based Reasoning: Techniques for Enterprise Systems*. Morgan Kaufmann Publishers Inc, San Francisco, CA.

Witten, I.H., Fran, E., Hall, M.A., Pal, C.J. (2016). *Data Mining: Practical Machine Learning Tools and Techniques*. Morgan Kaufmann Publishers Inc, San Francisco, CA.

List of Authors

Carlos J. ALONSO-GONZÁLEZ
Department of Computer Science
University of Valladolid
Spain

Anibal BREGON
Department of Computer Science
University of Valladolid
Spain

Erik FRISK
Department of Electrical Engineering
Linköping University
Sweden

Mattias KRYSANDER
Department of Electrical Engineering
Linköping University
Sweden

Gregory PROVAN
School of Computer Science and IT
University College Cork
Ireland

Vicenç PUIG
Institute for Robotics and
Industrial Informatics
Polytechnic University of Catalonia
Barcelona
Spain

Belarmino PULIDO
Department of Computer Science
University of Valladolid
Spain

Silvio SIMANI
Department of Engineering
University of Ferrara
Emilia-Romagna
Italy

Ye WANG
Department of Electrical and
Electronic Engineering
University of Melbourne
Australia

Index

A, B

analytical redundancy relation (ARR), 62, 72
bank of observers, 22
bi-adjacency matrix, 58

C, D

canonical decomposition, 67
conflicts, 211
diagnosability analysis, 64
diagnosis, 197, 210
differential constraint, 219
Dulmage–Mendelsohn decomposition, 60

E, F

exactly determined part, 61
fault detectability, 65
 analysis, 64
 active robustness, 35
 disturbance decoupling, 31
 H_∞ methods, 32, 34
 in dynamic systems, 1
 neural networks, 39
 passive robustness, 35
 performance index, 34
 structural, 65
fault detection and isolation (FDI)
 model-based, 2
Fault Diagnosis Toolbox, 58
 pattern recognition, 39
fault isolability, 65
 analysis, 64, 65
 matrix, 67
 structural, 66
fault location, 5
 actuator fault, 8
 multiplicative fault, 7
 sensor fault, 7
 sensor noise, 7
 transfer function model, 9
fuzzy clustering, 170
 c-means algorithm, 172
 Gustafson–Kessel algorithm, 174
 product–space, 176, 181
fuzzy model
 antecented fuzzy sets, 167, 168, 183
 consequent crisp functions, 167, 168, 185
 defuzzification, 169

G, H

Hall components, 62
Hankel matrix, 136

M

matching, 62
  maximal, 62
  perfect, 62
minimal
  hitting set, 77
  sensor set, 74
  structurally overdetermined (MSO) set, 69
  test equation support (MTES), 71
  test support (MTS), 71
model
  Auto Regressive eXogenous (ARX), 133
    nonlinear, 180
  fault location, 3, 8
  piecewise affine, 147
  Singleton, 169
  statespace, 9
  structural, 58
  Takagi–Sugeno, 167, 169
monitorable equation, 66
  algorithm, 71

N, O, P

neuro-fuzzy, 40
  B-spline, 42
  hierarchical networks, 43
  structure identification, 43
  Sugeno-type, 42
overdetermined, 62
parameter estimation
  equation error method, 15
  output error method, 17
parity equations, 24
proper structurally overdetermined (PSO) set, 69

R

reduced structure graph, 59
redundancy, 62
  degree, 69
regression
  nonlinear, 178
residual
  analysis, 2, 28, 29
    fuzzy decision-making, 38
  compared with threshold, 14
  evaluation, 3
    fuzzy threshold, 45
  generation, 3, 11, 13
    adaptive threshold, 36
    neuro-fuzzy, 44, 45
    via parameter estimation, 15
  generator, 72
  Kalman filters approach, 21
  multi-input multi-output (MIMO) processes, 22
  observer–based approach, 18
  output observers, 23
  parity equations, 26
  techniques, 14

S

sensor placement, 74
structural
  fault detectability, 65
  fault isolability, 66
  rank, 62
system
  affine, 131, 154
  Frisch scheme, 131
    algebraic case, 139
    dynamic case, 141
    multi-input multi-output (MIMO) case, 145
  fuzzy, 37, 164, 165
  identification, 131
  linear, 6, 133
  model-based fault detection and isolation (FDI), 2, 3
  nonlinear, 37, 146
system model
  error in variable (EIV), 8, 132, 133, 156
  fuzzy systems structure estimation, 177
  hybrid, 147
  state–space realization, 134

T, U

test equation support (TES), 70
test support (TS), 70, 71
testable submodels, 69
underdetermined part, 61

Summary of Volume 2

Chapter 1. Nonlinear Methods for Fault Diagnosis
Silvio SIMANI and Paolo CASTALDI

1.1. Introduction
1.2. Fault diagnosis tasks
1.2.1. Residual generation task
1.2.2. Residual evaluation task
1.3. Model-based fault diagnosis
1.3.1. Parity space relations
1.3.2. Observer-based approaches
1.3.3. Nonlinear filtering methods
1.3.4. Nonlinear geometric approach strategy
1.4. Data-driven fault diagnosis
1.4.1. Online identification methods
1.4.2. Machine learning approaches to fault diagnosis
1.5. Model-based and data-driven integrated fault diagnosis
1.6. Robust fault diagnosis problem
1.7. Summary
1.8. References

Chapter 2. Linear Parameter Varying Methods
Mickael RODRIGUES, Habib HAMDI and Didier THEILLIOL

2.1. Introduction
2.2. Preliminaries: a classical approach
2.3. Problem statement
2.4. Robust active fault-tolerant control design
2.4.1. Robust observer-based FTC design
2.4.2. Stability analysis
2.5. Application: an anaerobic bioreactor
2.6. Conclusion
2.7. References

Chapter 3. Fuzzy and Neural Network Approaches

Marcin WITCZAK, Marcin PAZERA, Norbert KUKUROWSKI and Marcin MRUGALSKI

3.1. Introduction
3.2. Fuzzy model design
3.2.1. Takagi–Sugeno systems
3.2.2. Generation of TS models via nonlinear embedding
3.3. Neural model design
3.3.1. Recurrent neural network
3.3.2. Identification of the neural model uncertainty
3.4. Fault estimation and diagnosis
3.4.1. Actuator fault estimation using neural networks
3.4.2. Sensor and actuator fault estimation using fuzzy logic
3.5. Fault-tolerant control
3.5.1. An overview of the fault-tolerant scheme
3.5.2. Robust fault estimation and control
3.5.3. Derivation of a robust invariant set
3.5.4. Efficient predictive FTC
3.6. Illustrative examples
3.6.1. Sensor and actuator fault estimation example
3.6.2. Fault-tolerant control example
3.7. Conclusion
3.8. Acknowledgment
3.9. References

Chapter 4. Model Predictive Control Methods

Krzysztof PATAN

4.1. Introduction
4.2. Idea of MPC
4.3. Robustness of MPC
4.4. Neural-network-based robust MPC
4.4.1. Neural network models
4.4.2. Nonlinear MPC
4.4.3. Approximate MPC
4.4.4. Robust nonlinear MPC
4.4.5. Robust approximate MPC
4.5. Robust control of a pneumatic servo
4.5.1. Robust nonlinear neural-network-based MPC
4.5.2. Robust approximate neural-network-based MPC
4.6. Conclusion
4.7. References

Chapter 5. Nonlinear Modeling for Fault-tolerant Control
Silvio SIMANI and Paolo CASTALDI

5.1. Introduction
5.1.1. Joint fault diagnosis and control
5.1.2. Nonlinear adaptive fault estimators
5.1.3. Fuzzy fault-tolerant control
5.1.4. Recursive adaptive control
5.1.5. Sustainable control
5.2. Fault-tolerant control strategies
5.2.1. Fault tolerance and compensation
5.3. Fault diagnosis and tolerant control
5.3.1. Fault-tolerant control design
5.4. Summary
5.5. References

Chapter 6. Virtual Sensors and Actuators
Damiano ROTONDO and Vicenç PUIG

6.1. Introduction
6.2. Problem statement
6.3. Virtual sensors and virtual actuators
6.4. LMI-based design
6.5. Additional considerations
6.6. Application example
6.6.1. Virtual actuator
6.6.2. Virtual sensors
6.7. Conclusion
6.8. References

Chapter 7. Conclusions
Vicenç PUIG and Silvio SIMANI

7.1. Introduction
7.2. Closing remarks
7.3. References

Chapter 8. Open Research Issues
Vicenç PUIG and Silvio SIMANI

8.1. Further works and open problems
8.1.1. Sustainable control design objectives
8.1.2. Sustainable control concepts and approaches
8.1.3. Sustainable control approaches and working methods
8.1.4. Sustainable control design ambition

8.1.5. Sustainable control innovation potentials
8.1.6. Sustainable control expected impacts
8.2. Summary
8.3. References

Printed and bound by CPI Group (UK) Ltd, Croydon, CR0 4YY